DEVELOPMENT

This volume collects essays from prominent intellectuals and public figures based on talks given at the 2015 Darwin College Lectures on the theme of 'development'. The writers are world-renowned experts in such diverse fields as architecture, astronomy, biology, climate science, economy, psychology, sports, and technology. *Development* includes contributions from developmental biologist and Nobel laureate John B. Gurdon, Olympic gold medallist Katherine Grainger, astronomer and cosmologist Richard Ellis, developmental psychologist Bruce Hood, former Met Office Chief Scientist Julia Slingo, architect Michael Pawlyn, development economist Ha-Joon Chang, and serial entrepreneur Hermann Hauser. While their perspectives and interpretations of development vary widely, their essays are linked by a common desire to describe and understand how things change, usually in the direction of ever-increasing complexity.

Written with the lay reader in mind, this interdisciplinary book is a must-read for anybody interested in the mechanisms underlying the changes we see in the world around us.

TORSTEN KRUDE is a University Senior Lecturer in Cell Biology at the University of Cambridge, and a Fellow at Darwin College. His research on the regulation of DNA replication in human cells has been widely published in leading international journals. In 2012 he received the prestigious Pilkington Prize for excellence in teaching by the University of Cambridge. He oversees the Darwin College Lecture Series and is also responsible for the development of the College Gardens.

SARA T. BAKER joined the Faculty of Education at the University of Cambridge in 2011. She performs basic research on cognitive development in the pre-school years, when children learn about a constantly changing world, and translates this research into educational contexts. Her projects have been funded by the Newton Trust, the ESRC, and the LEGO Foundation. She is a Deputy Dean at Darwin College and a member of the College Council.

THE DARWIN COLLEGE LECTURES

These essays are developed from the 2015 Darwin College Lecture Series. For more than 30 years now, these popular Cambridge talks have taken a single theme each year. Internationally distinguished scholars, skilled as popularizers, address the theme from the point of view of eight different arts and sciences disciplines.
Subjects covered in the series include:

2015 DEVELOPMENT
eds. Torsten Krude and Sara T. Baker
pb 9781108447379

2014 PLAGUES
eds. Jonathan L. Heeney and Sven Friedemann
pb 9781316644768

2013 FORESIGHT
eds. Lawrence W. Sherman and David Allan Feller
pb 9781107512368

2012 LIFE
eds. William Brown and Andrew Fabian
pb 9781107612556

2011 BEAUTY
eds. Lauren Arrington, Zoe Leinhardt and Philip Dawid
pb 9781107693432

2010 RISK
eds. Layla Skinns, Michael Scott and Tony Cox
pb 9780521171977

2009 DARWIN
eds. William Brown and Andrew C. Fabian
pb 9780521131957

2008 SERENDIPITY
eds. Mark de Rond and Iain Morley
pb 9780521181815

2007 IDENTITY
eds. Giselle Walker and Elisabeth Leedham-Green
pb 9780521897266

2006 SURVIVAL
ed. Emily Shuckburgh
pb 9780521710206

Development

Mechanisms of Change

Edited by *Torsten Krude* and *Sara T. Baker*

University of Cambridge

CAMBRIDGE
UNIVERSITY PRESS

University Printing House, Cambridge CB2 8BS, United Kingdom

One Liberty Plaza, 20th Floor, New York, NY 10006, USA

477 Williamstown Road, Port Melbourne, VIC 3207, Australia

314–321, 3rd Floor, Plot 3, Splendor Forum, Jasola District Centre, New Delhi – 110025, India

79 Anson Road, #06–04/06, Singapore 079906

Cambridge University Press is part of the University of Cambridge.

It furthers the University's mission by disseminating knowledge in the pursuit of
education, learning, and research at the highest international levels of excellence.

www.cambridge.org
Information on this title: www.cambridge.org/9781108447379
DOI: 10.1017/9781108686099

First published 2019

Printed and bound in Great Britain by Clays Ltd, Elcograf S.p.A.

A catalogue record for this publication is available from the British Library.

Library of Congress Cataloging-in-Publication Data
Names: Krude, Torsten, editor. | Baker, Sara T., 1978- editor.
Title: Development : mechanisms of change / edited by Torsten Krude
    (University of Cambridge), Sara T. Baker (University of Cambridge).
Description: Cambridge ; New York, NY : Cambridge University Press, 2019. |
    Series: The Darwin College lectures | Includes bibliographical references.
Identifiers: LCCN 2018027882 | ISBN 9781108447379 (pbk. : alk. paper)
Subjects: LCSH: Developmental biology. | Growth. | Change. | Age and intelligence. |
    Cosmology. | Climatology. | Economic development. | Sustainable development.
Classification: LCC QP83.8 .D48 2019 | DDC 612.6–dc23
LC record available at https://lccn.loc.gov/2018027882

ISBN 978-1-108-44737-9 Paperback

Contents

Figures

Notes on Contributors

John Gurdon Professor Sir John Bertrand Gurdon FRS is a developmental biologist, whose work has pioneered nuclear transfer and reprogramming, and led to the first cloning of a vertebrate organism.

John Gurdon was judged at school to be wholly unsuited to science, having come bottom in a class of 250 in Biology. However, he was able to take up science at Oxford University, where he also did a PhD. His work led to the concept that an egg has the ability to rejuvenate the nucleus of an adult cell, and hence to the current prospect of replacing aged and diseased cells in humans with new cells derived from other body cells such as skin. In mid-career he moved to Cambridge, where he still works in the University, in a major research institute that has been named in his honour, The Gurdon Institute. He is an Honorary Fellow of Christ Church, Oxford, and of Magdalene and Churchill Colleges in Cambridge. He served as Master of Magdalene College, Cambridge, from 1995–2002. He has received a number of awards, including the Copley Medal of the Royal Society (2003), the Lasker Award for Basic Medical Research (2009), and the Nobel Prize for Physiology or Medicine (2012).

Katherine Grainger Dame Katherine Grainger DBE is Britain's most successful international female rower and most-decorated female Olympian.

Originally from Glasgow, Katherine Grainger studied at Edinburgh University, where she took up the sport in 1993. Her international rowing career began in 1997, and since that time she has won six world championships and four Olympic silver medals (Sydney, Athens, Beijing, and Rio). In 2012 she won the Olympic gold medal with Anna Watkins in the double scull. Katherine Grainger is the only British female athlete from any sport to have won medals in five consecutive Olympic Games. She was awarded an MBE in 2006, a CBE for services to rowing in 2012, and a DBE for services to sport and charity in 2017. She obtained an Honours LLB from Edinburgh

University, an MPhil in Medical Law and Medical Ethics from Glasgow University, and a PhD in the Law of Homicide at King's College, London. In 2015, she became Chancellor of Oxford Brookes University. Alongside her studies, she wrote her autobiography *Dreams Do Come True*, which was published in June 2013.

Richard Ellis Professor Richard Salisbury Ellis CBE FRS is an astronomer and has charted the earliest period of cosmic history, when the first galaxies emerged.

A Welshman by birth, Richard Ellis was an undergraduate at University College London, and gained his PhD at Oxford. He became a professor of astronomy at the University of Durham in 1985. He was appointed the Plumian Professor at Cambridge in 1993, and served as Director of the Institute of Astronomy from 1994 to 1999. He then became the Steele Professor of Astronomy at the California Institute of Technology in Pasadena. Since 2015, he is Professor of Astrophysics at University College London. He has published widely on topics in observational cosmology and galaxy evolution. Within both the UK and the USA he has led discussions for new observational facilities, the most recent example being the Thirty Meter Telescope atop Mauna Kea on the island of Hawaii. Richard Ellis is a Fellow of the Royal Society and was awarded the Gruber Cosmology Prize for his part in the discovery of the accelerating Universe (2007), the Carl Sagan Memorial Prize (2017), and the Gold Medal of the Royal Astronomical Society (2011).

Bruce Hood Professor Bruce MacFarlane Hood is an experimental psychologist who researches early child development and specialises in the development of the mind.

Bruce Hood is Canadian-born, but moved to Scotland, where he was an undergraduate and obtained an MPhil at the University of Dundee. He then obtained his PhD from Cambridge and worked at MIT and Harvard before returning to the UK. Bruce Hood is currently the Professor of Developmental Psychology in Society at the University of Bristol, where he has worked for the past 16 years. He is the Director of the Bristol Cognitive Development Centre that researches early child development, and has published three popular science books on the development of the mind. In 2011, he presented the Royal Institution Christmas Lectures, 'Meet Your Brain', and received the Public Engagement and Media Awards from The British Psychological Society in 2013. He is a Fellow of the Association for Psychological Science, the Royal Society of Biology, and the Royal Institution of Great Britain.

Julia Slingo Professor Dame Julia Mary Slingo DBE FRS is a meteorologist and climate scientist.

Julia Slingo obtained her BSc and PhD at the University of Bristol and subsequently worked for the Met Office, the European Centre for Medium-Range Weather Forecasts in Reading, and the National Center for Atmospheric Research in the USA. She returned to the UK and became the Director of Climate Research in the NERC's National Centre for Atmospheric Science at the University of Reading. She became the first female Professor of Meteorology in the UK. In 2006, Julia Slingo founded the Walker Institute for Climate System Research at Reading, aimed at addressing the cross-disciplinary challenges of climate change and its impacts. She became the Met Office Chief Scientist in 2009, where she led a team on a portfolio of research that underpins weather forecasting, climate prediction, and climate change projections. Since joining the Met Office she has sought to integrate the UK community in weather and climate research to ensure that the UK receives maximum benefit from its science investments. She was awarded an OBE in 2008 for services to environmental and climate science and a DBE for services to weather and climate science in 2014.

Michael Pawlyn Michael Pawlyn is a London-based architect who has been described as a pioneer of biomimicry. He established the company Exploration Architecture to work on biomimetic architecture and the development of sustainable design.

After graduating from University College London and the University of Bath with a BSc and a BArch, respectively, Michael Pawlyn worked with Grimshaw Architects for 10 years and was central to the team that radically reinvented horticultural architecture for the Eden Project. He was responsible for leading the design of the Warm Temperate and Humid Tropics Biomes and the subsequent phases that included proposals for a third Biome for plants from dry tropical regions. He established the company Exploration Architecture in 2007 to focus exclusively on biomimicry, and in 2008 the company was short-listed for the Young Architect of the Year Award and the internationally renowned Buckminster Fuller Challenge. He is a Founding Partner of the Sahara Forest Project – a company established to deliver concrete initiatives for restorative growth. Michael Pawlyn has lectured widely on the subject of sustainable design in the UK and abroad, and delivered talks at the Royal Society of Arts, at Google's annual Zeitgeist conference and became one of only a small handful of architects to have a talk posted on TED.com in 2011. In the same year, his book *Biomimicry in Architecture* was published by the Royal Institute of British Architects.

Ha-Joon Chang Dr Ha-Joon Chang is an institutional economist who specialises in Development Economics.

Ha-Joon Chang graduated from the Department of Economics at Seoul National University in his native South Korea and subsequently moved to Cambridge, where he obtained his PhD in 1991. He currently teaches economics at the University of Cambridge and is a Reader in the Political Economy of Development. In addition to numerous journal articles and book chapters, he has published 15 authored books (four co-authored) and 10 edited books, including *The Political Economy of Industrial Policy, Kicking Away the Ladder, Bad Samaritans, 23 Things They Don't Tell You about Capitalism,* and *Economics: The User's Guide.* He has developed a critique of neo-liberal capitalism particularly in developing countries, and his writings remain accessible to the general public. His work has been translated and published in 36 languages and 39 countries. He is the winner of the Gunnar Myrdal Prize in 2003 and the Wassily Leontief Prize in 2005.

Hermann Hauser Dr Hermann Maria Hauser KBE FRS is a serial entrepreneur and co-founder of Amadeus Capital Partners.

Hermann Hauser obtained a Masters degree at Vienna University in his native Austria before moving to Cambridge, where he obtained a PhD in Physics. He has wide experience in developing and financing companies in the information technology sector. He co-founded a number of high-tech companies, including Acorn Computers (which spun out ARM), E-trade UK, Virata, and Cambridge Network. Subsequently he became Vice President of Research at Olivetti. During his tenure at Olivetti, he established a global network of research laboratories. Since leaving Olivetti, he has founded over 20 technology companies. In 1997, he co-founded Amadeus Capital Partners. At Amadeus he invested in CSR, Solexa, Icera, Xmos, and Cambridge Broadband. Hermann Hauser is a Fellow of the Royal Society, the Institute of Physics, and the Royal Academy of Engineering. He is an Honorary Fellow of Hughes Hall and King's College, Cambridge. In 2001, he was awarded an Honorary CBE for 'innovative service to the UK enterprise sector'. In 2004 he was made a member of the Government's Council for Science and Technology and in 2013 he was made a Distinguished Fellow of BCS, the Chartered Institute for IT. In 2015 he was awarded a KBE for services to engineering and industry.

Acknowledgments

The collection of essays presented in this volume, and the underlying Darwin College Lectures, would not have been possible without the help of many people. Torsten Krude and Sara T. Baker therefore wish to thank Professor Mary Fowler, Master of Darwin College, and the team of committed members of Darwin College who helped organise and facilitate the Lecture Series, the Syndics and editors at Cambridge University Press who enabled and supported publication of this volume, and, most notably, Janet Gibson for her outstanding expertise, commitment, and enthusiasm from start to finish.

Introduction

TORSTEN KRUDE AND SARA T. BAKER

For more than 30 years, Darwin College Cambridge has been hosting an annual series of eight public lectures taking place on subsequent Friday afternoons during winter and early spring. Each lecture series is assembled around a general theme that reaches out into the Arts and Humanities, the Natural and the Social Sciences, and beyond. In 2015, the theme was 'Development'. This volume contains a series of eight essays that are based on these eight lectures.

Why development? When tasked to define what development is, different ideas come forward from different disciplines. For instance, a biologist would think about the development of an organism from a fertilised egg, a process by which a living organism changes its shape and complexity over time. An athlete, craftsperson, or artist might see development as a personal process of acquiring skills over time through dedicated training and education. Parents would see development of their children as they grow and become unique and complex human personalities. Architects and civil engineers might see it as a process by which their work creates new buildings and structures that change cities and the people living in them. Economists, entrepreneurs, or people working in the finance sector would consider development in terms of the changes to wealth or financial assets over a period of time, often influenced by their ideas or actions. Therefore, the concept of development has many facets; it is interdisciplinary, but has an underlying common principle.

Development is a dynamic process by which things change over time, usually towards greater complexity. The underlying mechanisms bringing about these changes, however, are diverse and not always clear. They can include deterministic internal driving forces that result in predictable results. They can include external manipulations that shape and steer a particular process. Sometimes these two aspects operate side-by-side. The timescales over which we can observe a developmental process stretch over many orders of magnitude. On the one hand, many biological and social developments can be observed in real time during the lifetime of the observer. Astronomical developments, on the other hand, happen over billions of years, and their analysis requires reconstruction from observations made through space and time.

In order to understand development both in the human world and in the natural world, we need to elaborate a mechanistic understanding, in terms of the locus of change, the antecedents and consequents, the shape and the rate of change. We need to ask whether change is linear or non-linear. Are there significant discontinuities, where do qualitative changes happen, is there a clear before and after? Is there a value in development, will it lead to progress? Is it an autonomous process, or is it guided? Individual responses to these questions will be specific to the social and natural sciences. However, sometimes they reverberate back and forth to each other, in the same way as a discursive exchange between different disciplines may be observed in the dining hall over a meal at Darwin College.

Conceptually, development therefore provides an exciting topic for wide and interdisciplinary exploration. When organising the 2015 Darwin lecture series, we brought together eight lecturers from a broad range of disciplines, who approached the topic from their specific vantage points. The following eight chapters in this volume are based on these lectures.

In the first chapter, **John Gurdon** illustrates the fascinating world of developmental biology, and describes the ground-breaking discoveries that led to his Nobel Prize in Physiology or Medicine in 2012. He explores basic mechanisms of animal development first, by which a fertilised egg of a frog turns itself into a tadpole and eventually into a frog, without guidance or support from another frog. He then turns to human experimental intervention in this process and describes how, for instance, the nucleus from a skin cell of an adult frog can be repro-grammed in the laboratory by exposure to the unique environment of the egg, and subsequently develop into a new frog. A consequence of this development is the technological possibility of replacing old and diseased cells in an adult organism by reprogrammed and rejuvenated adult cells, paving the way for cell replacement therapies. The narrative of this chapter is that the autonomous process of animal development can be reprogrammed by human intervention, which in itself constitutes a technological development.

When it comes to the second chapter, **Katherine Grainger** exemplifies the development of an outstanding athlete. She shares an autobiographical account as an Olympian who saw British sport, particularly rowing, develop dramatically during her lifetime. She explores what forces contribute to, or hinder, an athlete's performance. In her experience, the concept of marginal gains was very powerful in achieving successful leaps forward at the highest levels of competition. Her personal story is interwoven with a commentary on the development of a national movement and the multitude of causal agents coming together to support progress, and ultimately to produce individual and collective successes at the 2012 London Olympics.

The third chapter takes an astronomer's perspective on the question of development. **Richard Ellis** explores the development of galaxies and asks how it is possible to know our cosmic history, from our current vantage point looking backwards in time to when our Universe was just about 5% of its present age. Observation of galaxies of different ages, he argues, holds information about their formation and the evolution of the Universe as a whole. The challenge we face is how to reconstruct this story given the technological means at our disposal. Here we see that the development of new technologies on this planet during modern times has enabled us to study the development of galaxies over a cosmic timescale.

The volume's fourth chapter moves into developmental psychology and explores what gives rise to the human experience of a sense of self. **Bruce Hood** argues that we develop a sense of self as a consequence of striving for a coherent picture of ourselves and the world around us, even if the picture is a mere illusion. The so-called 'self-illusion' arises in the face of a multitude of influences, factors, and experiences during a lifetime, while early developmental experiences hold a critical importance.

In the fifth chapter, on the development of climate science, **Julia Slingo** explains how advances in human capacity for data capture and data analysis are tightly linked with advances in our understanding of the dynamic climate around us. Her chapter focuses on the development of climate science as a scientific discipline with an enormous impact. Julia Slingo offers her personal perspective as a lead climatologist on how climate science has emerged. It is rooted in meteorology and oceanography, but has now become a means of understanding 'how our climate system works and why climate change is arguably one of the greatest challenges facing us in the twenty-first century'.

The architect **Michael Pawlyn** writes in the sixth chapter about the development of sustainable design. He explores the concept of biomimicry as a design tool, which may become instrumental in shifting mankind from the industrial to an ecological age. Biomimicry uses adaptations in the biological world, which arose during 3.8 billon years of evolution on this planet, as an inspiration to inform and guide sustainable design. This chapter presents examples of the development of designs for office buildings, a data centre, and cityscapes, all based on biomimicry. It concludes with the ambitious Sahara Forest Project as a means to grow crops in desert areas, to generate clean energy, and to reverse desertification. Application of biomimicry, he argues, has the potential for the development of sustainable human habitats.

The seventh chapter in the volume focuses on economic development. **Ha-Joon Chang** calls for a shift in how we describe the mechanisms underlying a country's economic development process. Rather than focusing exclusively on poverty reduction as the main driving force, he argues that production should be viewed as being of central and fundamental importance, with contributions of technological learning and individual developments. His chapter considers key questions about development, such as the extent to which economic development and long-term growth are related, and whether development in financial terms is linked to development in terms of human capital.

The volume concludes with a chapter on technology development by the entrepreneur **Hermann Hauser**. New technologies have already been identified as driving factors of development in several of the preceding chapters. In this final chapter, Hermann Hauser takes this concept further and focuses on computation and information technology as a result of human activity. He explores the development of computing in multiple waves, arguing that there are qualitative changes marking the transition from one to the next. Looking forward, he discusses the development and future implications of machine learning and artificial intelligence for human society, thus concluding his chapter and the discussion of development collected in this volume.

The 2015 Darwin Lecture Series on Development also coincided with the coming of age of the eight United Nations Millennium Development Goals. They comprise a set of priorities agreed by the international community at the Millennium Summit in 2000, notably for reducing poverty and increasing access to education in so-called developing countries. In 2015, the United Nations extended these to the newly minted 17 Sustainable Development Goals, which are to be attained by 2030. These new goals are more inclusive: instead of targeting only so-called developing countries, the new goals apply to all countries around the globe. The new goals are also more integrative: instead of focusing on quantifiers like how many people are completing primary education, there is now also a focus on quality. We can therefore take the opportunity now, on the threshold of these new global monitoring milestones, to reflect on what the chapters in the present volume on development suggest about our future trajectory. Where will we be in 2030, with respect to artificial intelligence, understanding ourselves and our galaxy, stem cell therapy, human athletic feats, the world economy, climate science, sustainability, and harmony with nature? We might look back to 2015 and contemplate new developments that will have led to a changed world, which lies ahead of us.

1 Animal Development and Reprogramming

JOHN B. GURDON

We all start life as a single fertilised egg cell, but that single cell is then converted into the highly complex organism that we all are. In the first part of this essay, I will address how that happens. In the second part, I will discuss how we can make this process go backwards, and eventually provide new rejuvenated cells when our own deteriorate with age or disease. I will also address the question of whether we really can replace old cells with new cells, and, if so, to which extent. I will finally address the ethical and legal constraints of making this procedure more generally available.

From Egg to Adult: How Does It Happen?

In most species, excluding mammals and humans, the very early stages of development can be seen. During the development of a frog, a fertilised egg first develops synchronously, and within only a few hours the egg has turned itself into a ball of several thousand cells. Soon after, some of the cells on the outside start to move to the inside of the ball and, within a day or two, the future brain, nervous system, and other parts of the embryo can be seen to form. In mammals, early development takes place in the mother, and the mother's cells help to guide the embryo as it forms. But in all other animals early development takes place entirely independently of the mother. The different kinds of cells of which we consist appear progressively. At an early stage new embryo cells take a decision whether to go in the direction of brain and skin, muscle or intestine. Later, those that have gone in the first direction take another decision and follow separate pathways to reach their eventual fate (Figure 1.1). Once a cell has embarked on a pathway leading to a particular fate, it and

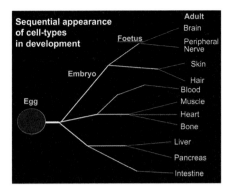

FIGURE 1.1 From egg to adult. As embryo cells divide, they and their daughter cells become progressively restricted in terms of the kinds of cells that they can form.

its daughter cells do not change or go backwards. Therefore, as an embryo grows and differentiates, its cells become progressively committed to particular pathways, leading to specific cell types.

How Does the Embryo Know What to Do?

We are familiar with frogspawn in a pond. Initially, each egg is a single cell, and in a few days each one has turned itself into a swimming tadpole. How does it know how to do this? The parent frogs have long since disappeared, and give no guidance to the eggs on how to develop.

Before powerful microscopes were invented, it was thought that each egg or sperm might have a little creature inside it, dubbed a homunculus, and it just had to grow (Figure 1.2). Later, in the nineteenth and early twentieth centuries, the very influential German embryologist Ernst Haeckel (1834–1919) was impressed by how, in early life, the embryos of many different kinds of animals look similar. He made the proposition that 'phylogeny is the mechanical cause of ontogeny'. Many people revered Haeckel's global view of development, but the phrase I have quoted did not give any meaningful explanation of how development works. At the time, Haeckel's position as a very senior German professor led others to accept his pronouncements without questioning them. However, later commentators treated him often as what, I think, we call

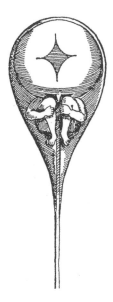

FIGURE 1.2 An early idea that a human sperm might contain a miniature man, a homunculus, which could grow to an adult once in the fertilised egg. (From Hartsoeker (1694).)

the 'whipping boy'. When something did not make sense, they blamed it on Haeckel. For example, to quote one very highly regarded later commentator: 'Haeckel's greatest disservice was not his total ignorance of exception to his rule, but his emphasis on his irrefutable explanation of the mechanical cause of development. He thereby distracted those who might otherwise have made a valuable contribution to this whole field.' (Hamburger, 1988).

But what have we learned about the principles of development, what can we say now about how an egg can turn itself into a complete organism? If we look at the inside of a frog egg, there is no indication of any kind of organism, or type of cell, inside. All we can see is a tiny nucleus at the top in a huge mass of the so-called cytoplasm, most of which is yolk platelets, a source of nutrition for the embryo (Figure 1.3). We now know that there are two fundamental mechanisms by which this mass of apparently disorganised cytoplasm turns itself into cells of different kinds.

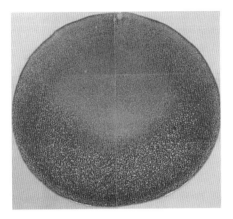

FIGURE 1.3 The unfertilised egg of a frog. Most of the egg consists of yolk (red), and apparently structureless cytoplasm (blue) in the middle. (Plate 8 of Hausen and Riebesell (1991). With permission of Springer Nature.)

Asymmetric Distribution of Parental Molecules

As long ago as 1905, Conklin made a detailed description of the marine mollusc Styela. He could see that the yellow-coloured cytoplasm of the undivided egg gradually became localised in one part of the egg, and subsequently in some of the cells derived from that region of the egg (Figure 1.4). This yellow material eventually became part of the muscle of the embryo. The basic concept that resulted from this observation is that the undivided egg has various substances in its cytoplasm and that these are gradually distributed in the early embryo so that each of them goes to different parts of the future embryo. The yellow pigmented material happens to mark those substances that become muscle, though the yellow pigment is itself only indicative of other muscle-forming substances (Figure 1.4). As development proceeds, these formative substances, originally present in the undivided egg, gradually become localised to those parts of the early embryo which will turn into various kinds of cells. Various external influences cause these substances to move to where they need to be. These include the exact position in which the sperm enters the egg, and the movement of substances under the influence of gravity. As a result of the progressive localisation of these formative substances, cells

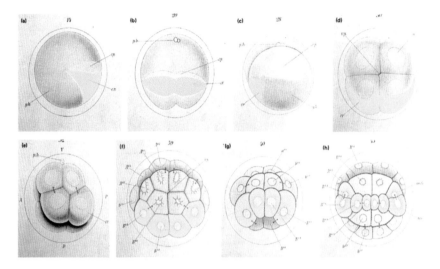

FIGURE 1.4 Localised substances in the early fertilised egg become progressively localised to the part of the embryo that will form a particular kind of cell. The muscle-forming substances of the egg of the mollusc are occupied by a yellow pigment. (From Conklin (1905).)

gradually become committed to their eventual fate, and alternative pathways are eliminated (Figure 1.1).

Signalling between Different Cells

We now know that cells of one kind in one part of an embryo send signals, in the form of molecules, to other cells elsewhere in the embryo. In the frog embryo, for example, the earliest distinction is between those cells at the top end of the embryo and those at the bottom (Figure 1.5). Substances which originated at the top end of the undivided egg, seen in yellow, become localised in the cells which will eventually form skin and brain, whereas substances at the other end of the egg, seen in blue, are localised in cells which will form the intestine and internal organs. After only a few hours, the blue cells send signal molecules upwards to the nearby yellow cells and cause the yellow cells to change their fate by making different substances, seen in green. These commit those cells to

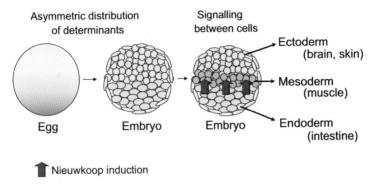

Asymmetric distribution of determinants

Signalling between cells

Egg → Embryo → Embryo

Ectoderm (brain, skin)

Mesoderm (muscle)

Endoderm (intestine)

Nieuwkoop induction

F I G U R E 1.5 The principle of signalling between cells. The yellow and blue substances are asymmetrically distributed to different poles of the egg, and then to different cells of the early embryo. When many cells have been formed, the 'blue' cells (the endoderm, giving rise to the intestine etc.) send molecular signals to the early 'yellow' cells (the ectoderm, giving rise to skin, brain etc.), causing them to change direction and become 'green' cells (the mesoderm, giving rise to muscle, bones, etc.).

become different parts of the body such as muscle, bone, etc. (Figure 1.5). This process of signalling between cells in one position and their neighbours elsewhere continues through development, and in later life.

The same principle lies behind the fundamental process of how stem cells operate. For example, deep in our skin are skin stem cells. As seen in Figure 1.6, the stem cell gradually accumulates skin-forming substances, and these are asymmetrically distributed as the stem cell grows and eventually divides (in the same way as this happens in eggs). The stem cell divides, typically asymmetrically, into one small daughter cell and another larger daughter cell. The small cell has accumulated a high concentration of these skin-forming substances, which cause that cell to activate genes resulting in the formation of skin. This process continues and repeats itself as the larger daughter cell, which remains a stem cell, undergoes subsequent divisions (Figure 1.6).

Very often the process of signalling works in a concentration-dependent way. As the signalling molecules spread across a cell or a group of cells from one part of the embryo, they become dilute as they spread further away. Consequently, cells near the source receive a high concentration of this substance and those further away a lower concentration. Amazingly, cells are able to sense the concentration of substance

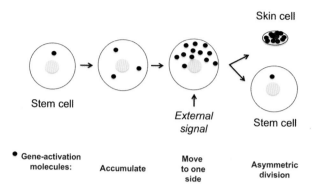

FIGURE 1.6 The principle of stem cell function. A stem cell makes skin-forming substances, which then become highly concentrated at one side of this cell (these gene-activating molecules are symbolised by black dots in the figure). After asymmetric cell division, they are present in the smaller daughter cell, which forms a skin cell. The larger daughter cell continues life as another stem cell.

that they receive, and respond accordingly. In this way, many different kinds of cell-types are formed by the concentration-dependent effects of one molecule along the concentration gradient of this molecule.

In summary, the way a single undivided egg cell turns itself into a complex embryo depends on two processes. One is the asymmetric distribution of cell-type-forming substances, and the other is the signalling from one kind of cell to other cells nearby. These two fundamental mechanisms enable an egg to turn itself progressively into an embryo, and into the many different kinds of cells that there are in the body.

Can Cells Be Made to Go Backwards in Development?

I have argued above that the process of normal development is a one-way process as cells gradually become more restricted in terms of what they can form; they and their daughter cells do not go backwards. This results in very stable cell-types of different kinds. Fortunately, we therefore do not find brain cells in our intestine or muscle cells in our skin. However, it is now possible, by experimental means, to make cells go backwards in life, and so to reverse this normally stable process of cell differentiation.

We need now to go back some 50 years to the 1950s to trace this technological development from the time when I first started as a graduate student. At that time it was not known whether all the different kinds of cells of our body have the same genes. A perfectly plausible idea at the time was that, as the embryo forms brain, skin, muscle, and all the other tissues, the genes needed for brain, for example, are lost from those cells that will form muscle.

The pioneering work of Briggs and King in 1952 succeeded in the technical achievement of moving the nucleus out of one kind of cell into an egg which itself had not been fertilised and had no nucleus. They were able to take the nucleus from an embryo cell, put it into the cytoplasm of an egg, and grow a normal tadpole. This opened the way to the process now called nuclear transplantation. My own involvement also started in the 1950s by doing these experiments with the South African frog *Xenopus laevis*, which was particularly favourable for this kind of work. I eventually found that it was possible to take the nucleus out of a functional intestine cell and implant that into an unfertilised egg whose own nucleus had been removed. As a result, it was possible to obtain an entirely normal adult, sexually mature frog from the combination of the nucleus of an intestine cell with an egg. This proved decisively that, as cells differentiate, in this case into intestine, they do not lose any genes needed for all the other kinds of cells that compose an adult animal. This gave rise to the fundamental conclusion that, as cells undergo specialisation into different tissue types, all genes of the body are retained in all cells. This makes it possible, in principle, to generate embryo cells from the nucleus of a skin cell, and hence to generate all the other kinds of cells that embryos normally give rise to. Thus the normally stable process of specialisation into intestine or other cell-types can be reversed experimentally by taking the nucleus from a specialised cell and transplanting it to an egg. The egg then forms the whole range of other kinds of cells that make up the body. In effect, the intestine cell nucleus is rejuvenated and made to go back to the beginning of life and start all over again (Figure 1.7).

There were all sorts of technical difficulties in achieving this basic experiment. First of all, the species of frog I used had eggs covered with an intensely elastic jelly, which made them impenetrable by any kind of needle that might be used to transplant a nucleus into the egg. Eventually this problem was solved by finding an appropriate wavelength of ultra-violet light that softened the jelly, and, by good luck, killed the nucleus of

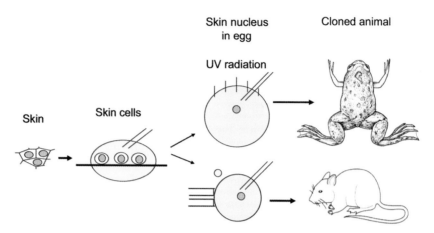

FIGURE 1.7 The principle of a nuclear transplantation experiment. The nucleus of an adult skin cell (pink) is transplanted into an egg (light blue), whose own chromosomes have been killed by ultraviolet irradiation (top), or removed from the egg (bottom). In some cases, the resulting embryo can become an adult cloned animal.

the egg. The next problem was to find a way of decisively marking the nucleus of the intestine cell so that it could be proved that the resulting adult frog did indeed come from the intestine cell nucleus and not from a failure to remove the nucleus of the egg. At that time, my PhD supervisor had a student who was struggling with a difficult problem and obtained an inexplicable result concerning the number of nucleoli in a nucleus in these frog cells. Dr Michael Fischberg, my supervisor, had the extraordinary wisdom not to ask the student to repeat all the experiments with new cells and new reagents. Instead, he said 'Please go back and find the actual animal which gave you these inexplicable results.' It turned out that a particular mutation had occurred, which altered the number of nucleoli in a cell. This was a natural genetic mutation, which turned out to be extremely valuable in enabling us to mark the descendants of a transplanted nucleus, as opposed to nuclei from an egg whose own nucleus had not been successfully removed. The adult frogs resulting from transplanted intestine nuclei eventually became males and females, and they were reproductively fertile. This proved that every single cell-type, including eggs and sperm, can be derived from the nucleus of an intestine cell.

Some decades later, the cloning of Dolly the sheep was announced. This was the first successful application of this nuclear transfer technique to mammals. Dolly has become very famous. Some people asked why did we not give names to our intestine-derived adult frogs. The answer was that we had so many of them that this would not have been realistic. Dolly the sheep was from the only successful nuclear transplant at that time, though subsequently nuclear transfer in mammals has been successful in a whole range of different species, even including humans.

Embryonic Stem Cells

The next major advance in this field, going from nuclear transfer to the eventual possibility of cell replacement, depended upon the discovery of embryonic stem cells by Martin Evans. He discovered, contrary to expectation, that it is possible to take cells from an early mouse embryo and place them in culture in such a way that these cells go on dividing indefinitely but nevertheless may remain embryonic. Amazingly, these permanent embryonic stem cells, as they are called, can be made to specialise when required. Special cell-type-forming substances, of the kind I have mentioned above when discussing signalling, can be added to these embryonic stem cells, and they can be made to form all different kinds of cells, including those of brain, heart, and other tissues. This extraordinary discovery, most appropriately awarded a Nobel Prize in 2006, has led to the current great interest in cell replacement therapy. The critical discovery made by these experiments of Martin Evans can be summarised as follows.

1. The cells can be grown from the early embryos of mice and humans, and are known as embryonic stem cells.
2. They can be grown indefinitely in the laboratory to make billions of unspecialised stem cells.
3. They can be made, at any time, to differentiate into heart, brain, and all other kinds of cells.

Induced Pluripotent Stem Cells

The next key advance in this field, leading to cell replacement therapy, was announced by Takahashi and Yamanaka (2006). As a result of this work, Yamanaka received a Nobel Prize for Physiology or Medicine in

2012. Takahashi and Yamanaka found that they could place cells from adult organs into culture in the laboratory. They identified four individual genes, which could be added, as pure DNA, to these cells. Very occasionally one of these cells would take up the combination of these four genes and flip from an adult state into an embryonic stem cell state. The fraction of all the cells treated that switched to an embryonic state was extremely small, about one in 10,000 or less. This work nevertheless established that it is possible to make development go backwards, as in the case of nuclear transfer, but, importantly, not requiring the use of eggs. Performing nuclear transfer experiments of the kind discussed above in humans would be unrealistic because it is extremely difficult to get a sufficient number of human eggs for this kind of work. If, however, the ultimate result of switching adult body cells back into embryo cells can be achieved, the provision of eggs is not required. This discovery led to a tidal wave of activity, including the creation of new institutes of stem cell biology all round the world at an amazing rate.

These induced pluripotent stem cells, as they are called, can now be derived from all sorts of easily accessible cells of adults. For example, skin cells can be used, as can blood cells. The resulting induced pluripotent stem cells can rather easily be switched into different fates, including heart, muscle, and, especially, brain cells. The substances, or molecules, which cause these cells to follow a desired fate are the same ones as those used during embryo development. These are the signalling molecules, which cause a cell to undergo the next specialisation step. For example, they are needed to cause the early embryo cells from one part of the embryo to become nerve or brain, as opposed to skin. This means that, once an adult cell has been switched back to an embryonic state, it can be pushed to form almost any desired other cell-type. Even though we do not know exactly how this works, it is very effective. For instance, various kinds of brain cells can be derived from skin using human cells. Likewise, beating heart muscle cells can be made in the laboratory, starting with skin cells from a monkey or other animals.

Prospects for Cell Replacement Therapy

The question now arises as to whether these new kinds of embryonic cells, or rejuvenated adult cells, can be used to treat patients. Can the derived cells be successfully transplanted back into patients in order to give therapeutic benefit? There is one area in which there seems to be a real prospect. This concerns the fairly common disorder known as age-related macular degeneration. In older adult humans, this is a fairly common disorder by which the photoreceptor eye cells die and central vision is lost, leading eventually to blindness.

Several laboratories and companies have been very active in generating the so-called pigmented retinal epithelium cells, which are required, in the eye, to keep the photoreceptors in a healthy state. It is now possible to grow these retinal pigmented epithelium cells from skin or other accessible adult cell-types. These cells can be transplanted to the back of the eye in humans, and they serve to replace the defective pigmented retinal epithelium cells, and so give support to the crucial photoreceptor cells. This procedure has been successful in animals and is now being tested in humans. The prospect is that this form of cell replacement therapy can arrest further deterioration of photoreceptor cells and hence preserve vision. It may even be possible to reverse the process of deterioration. The reason why this particular use of cell replacement seems to be so hopeful and successful is that a relatively small number of cells is required. A human consists of 10^{14} cells, and a major heart failure would result in the death of 10^{11} cells. For relief from age-related macular degeneration, only about 10,000–100,000 cells are needed. All these cells are of one type, and they can be grown in the laboratory. This is therefore the ideal situation in which to try to make use of rejuvenated adult cells for patient benefit.

Why is this treatment not yet available to patients? The problem is that, according to the laws of this and some other countries, no medical treatment is allowed to take place until it has been exhaustively tested and approved by so-called regulatory bodies. These bodies or committees are notoriously conservative, and may demand several years of work, tests, and further experiments before they will allow a new treatment to be made available to patients (Lachmann, 2012). In my view, the reason why these regulatory bodies or committees are so conservative is that they are intimidated by the legal profession. If anything were to go wrong, lawyers are able to award punitive damages against hospitals, doctors, and others. Even if a patient says that they are willing to sign away their legal rights to prosecute a doctor or hospital, members of the legal profession are able to refuse to accept that such a decision by the patient was made from a position of informed consent. It is perfectly straightforward to explain to a patient what the procedure involves and to tell the patient about any risk that might arise. Lawyers will argue whether such a patient has given 'informed consent'. In practice it is impossible to prove informed consent, therefore this great technology of cell replacement is likely still to be some years away from availability to patients. In my view, it is completely inappropriate for this restriction to exist. A patient should, in my view, be able to say that they have understood the procedure and they have chosen to have the replacement therapy applied to them. Particularly for elderly patients, the prospect of relief from macular degeneration and hence from the prospect of blindness is immensely attractive. It seems to me a tragedy that this wonderful technology is being withheld from willing patients for what are essentially administrative reasons.

Conclusions

We have traced over the last half century how the process of development and cell specialisation can be reversed by nuclear transfer and induced pluripotency techniques. It is now possible to derive functional adult cells of several different kinds from the rejuvenated embryonic cells, which themselves are derived from other adult cells. This procedure can be made to work, thereby enabling an individual to receive rejuvenated cells without the prospect of immunological rejection. I believe it is only a question of time before this technology will become applicable not just to relief from visual deterioration, but also for some other human disorders. It is not, however, likely that serious neurodegenerative diseases such as Alzheimer's and Parkinson's will be soon relieved by this technology. This is because the brain is very complicated, and many different kinds of cells seem to have to function cooperatively to relieve these conditions.

Nevertheless, the last half century has seen enormous advances in this field of understanding normal development and of making use of it for human therapy. If we can overcome administrative obstructions, I believe the prospects for cell replacement therapy are very promising.

Select Bibliography

Conklin, E. G. (1905). The organization and cell lineage of the ascidian egg. *Journal of the Academy of Natural Sciences of Philadelphia* 13, 1–119.

Gurdon, J. B. (2006). From nuclear transfer to nuclear reprogramming: the reversal of cell differentiation. *Annual Review of Cell and Developmental Biology*, 22, 1–22.

Hamburger, V. (1988). *The Heritage of Experimental Embryology: Hans Spemann and the Organizer.* Oxford: Oxford University Press.

Hartsoeker, N. (1694). *Essay de dioptrique.* Paris: Jean Anisson.

Hausen, P., & Riebesell, M. (1991). *The Early Development of Xenopus laevis: An Atlas of the Histology.* Berlin: Springer-Verlag.

Lachmann, P. J. (2012). The penumbra of thalidomide, the litigation culture and the licensing of pharmaceuticals. *Quarterly Journal of Medicine*, 105, 1179–1189.

Saxén, L., & Toivonen, S. (1962). *Primary Embryonic Induction.* London: Logo.

Takahashi, K., & Yamanaka, S. (2006). Induction of pluripotent stem cells from mouse embryonic and adult fibroblast cultures by defined factors. *Cell* 126, 663–676.

2 Development of an Athlete

KATHERINE GRAINGER

In this essay, which follows the Darwin lecture I gave in January 2015 in Cambridge, I will give a personal account of my own career in rowing and describe how I developed as an athlete and how rowing has developed as a highly competitive sport over the course of four Olympic Games.

The Beginnings at University

I did not expect rowing to be my life, or to be my career. It has been the most rewarding thing I have ever done and could have ever dreamed of doing. I did not start rowing until I went to university in Edinburgh to do law. If I am absolutely honest, this was based on the fact that I watched a lot of *LA Law* and other legal dramas and it felt that was what law was. I wanted to wear smart suits with shoulder pads and make impressive speeches about justice and save the world. But then I went to my first tax lecture and it dawned on me that it is not all like it is on TV.

It was at Edinburgh that I, by complete accident, fell into rowing. I had tried it once when I was growing up in Glasgow, but it was not anything my family did and it was not anything I knew much about. I remember when I went to university people said whatever you do, just try things. They said that being at university is the best time to meet new people, experience new places, and join new clubs. Therefore I went to the Freshers' Fair and joined up to as much as I could. I still, genuinely, have my membership card for the juggling club.

The rowing club was not one of the clubs to tempt me. Anyway, at one point I was alone, and being a fresher you know you do not want to be seen alone. It is that culling the weak person syndrome: you can't be seen on your own away from the herd because then you look as if you have got no friends and you will maybe never get any friends. Then I saw someone I recognised, so I went over to join her because then we were friends, stronger in a pack. She wanted to speak to the rowing club, and so I was hanging around on the sidelines waiting for her. Someone from the rowing club came over to me and said 'Do you want to hear about rowing?' I said 'no'. Rowing is a strange sport. Back then, I did not know anything about it. I was not interested. I had signed up to the juggling club and the abseiling club and the ski-ing and the climbing and the sailing and you name it. I was going to do all this outdoor stuff. Then she said 'Oh, I think you'd be really interested. I think you should at least have a go.' I said 'No. Too busy.' 'Oh no, but I think you would be really good.' I started thinking, 'Well, how good do you think I could be?' And then she said – this was the clincher – 'Well, come along on Thursday. We are going to have a meeting about it. You will hear all about it and then there is a free drink afterwards.' I thought 'Ok, yeah.' Well, I was a student. What can I say?

I went along with an air of reluctance but with an open mind. It was in a lecture theatre; I stood right at the back, detached but listening. They announced they only wanted 16 novice women but 52 had signed up. I remember thinking that I did not particularly want to row, but I did want to see if I could be one of the 16. And it was possibly then that I realised just how competitive I was.

I made it into the 16 and, I guess, that is where my career began. Initially it was not the sport that hooked me; it was the people involved. Those who are lucky enough to be rowers will understand this: the sport itself is quite addictive, but it is the people who matter. Somehow the people who are involved in rowing are special and incredible, crazy, insane, wacky characters, just brilliant people; really motivated, driven, eccentric, and batty, yes, but just wonderful charismatic people. Right through my university years, right up to being in the Olympic Team, I have met person after person through rowing who are exceptional individuals. That is what hooked me, that first year. The people absolutely hooked me and then, undoubtedly, I fell in love with the sport itself and ever since then I have been trying to be as good an athlete as I can be.

Rowing is something that I have now done for over 20 years, and I am still trying to perfect the rowing stroke. It is a stroke you do over and over and over again. We do it hundreds and thousands of times a day, and I still have not got it right. This is the attraction of the sport, because you are constantly developing as a person, as an athlete, as a technical model, as a physical model, a psychological model. You develop constantly and it is incredibly hard, but you get challenged every single day and you relish the challenge.

The Sydney 2000 Olympics

No one is born successful. No one is born with a gold medal. The whole thing is a huge development process. A massive journey, with no guarantees. It has been a big journey for me as an individual but also for the sport of rowing as well. In the 1996 Olympics, Steve Redgrave and Matthew Pinsent won the only gold medal in the whole British Team. Britain was 36th in the medal table with one gold medal, and it was seen as a disappointment all round. After that point a huge amount of money was put into British sport. Funding came in from the National Lottery from 1997 onwards, and it helped to transform sport. In rowing, for example, for the first time the women's team had a professional full-time coach, contracted up to the Sydney Olympics. It created long-term stability and vision.

I did not know the rowing world before funding began, but I was training with women who had massive debt, had large overdrafts as well as part-time jobs. It was not a professional set-up. The new idea was trying to professionalise things and to get athletes who could train full time and to see how competitive they then could be. I joined the team in 1997, and in the run up to the Sydney Olympics in 2000 I was in a quad. I was young and naïve and it was a lovely time. I was very ambitious and had massive confidence in the team I was part of, but this team had never won any Olympic medal in women's rowing in the history of the Olympics. No medal of any colour.

Therefore we came into Sydney knowing that, if one of our crews could break through that barrier, then we could start changing things. When you are in that development stage, there is no clear path to follow, there is nothing that has gone before to show the way. The men's team had been very successful, but the women's team just had not had consistent, sustained success. It was a question of trial and error in the lead up to the Sydney Games. And the biggest thing for me in Sydney, the strongest memory of that whole time, was just how much it was a living dream. It was simply a dream that we cared so much about, we were so passionate about. We were not basing it on facts and figures, and data and past results, because we did not have those. It was more on hopes and beliefs and a willingness to seize the opportunity.

The dream did not always go smoothly. An injury to one of the girls before the Olympics meant we had a new person coming into the boat and we went to the Games never having raced together before. However, the night before the final we had this incredible meeting where we simply talked to each other about what that race would mean to us and what we were trying to do and why. We really tapped into the emotional side of things and talked in a very open way that we had probably never done before, about our motivations and our hopes and our dreams. It absolutely united us; united us in a way we had not been in the time leading up to it.

We went out to race in the morning in the Sydney final just knowing that we had one chance. Every boat starts level on the start line. It doesn't matter how many Olympics you have been to, or how many medals you have won. Everyone starts level and everyone has got this 2,000 metres race. What matters is what you do in those 2,000 metres. That is the chance, the opportunity, every athlete is given.

We came together and agreed we would put everything out there on the line. We would give everything to each other for those 2,000 metres: no matter what pain we went through, we would give it to each other for this one race, probably the only race we would ever do together. It was incredibly powerful, and I remember us all talking about going into that dark tunnel that you get to in racing, where everything hurts and everything wants you to stop and how we knew that we would stay in that tunnel until we came out into the lightness at the end. I was the youngest and least experienced and was terrified of letting people down. It was my overriding fear. I would do anything and everything to be sure I was not the weak link in the race.

In the final, Ukraine was probably our biggest impediment to getting a medal, and I remember most of the race we were ahead of them. So I knew we were in medal contention. We had this incredible detailed structured race plan that we had discussed and planned and sorted and finalised leading up to the race, and we unfolded it bit by bit down the whole course, 250 metres at a time, disciplined, unpacking it and putting it all out there. Then, in the last part of the race you hear the roar of the crowds, you know the finish line is coming, you know this is your moment and we became simply four women screaming at each other rushing towards the finishing line. It was just uncontrolled passion. I don't even know what we were saying but we burst across the line. I knew we were ahead of Ukraine. I knew we'd got our medal. Huge celebrations.

We came into the landing stage and I was absolutely convinced we had won this bronze medal. There were hugs and tears of joy. Then we were told they could not do the medal ceremony because there was a photo finish. I assumed the photo finish was between the gold and silver. They said it was between silver and bronze: ourselves and the Russians. I had been so focused in my tunnel I had not even noticed there was another crew in the picture. My genuine reaction to that was 'Don't worry about the photo finish; we have won the Olympic medal and it doesn't matter what colour they come in.' And back then, I genuinely did not care what colour of medal came. The medal was everything, the medal was the big thing. That shows how things change, as I would have been happy with any colour on that day.

However, I can tell you how fast your expectations change, because they took 10 minutes to decide the photo finish. It was genuinely the smallest margin they could tell between the two boats. At the beginning I was over the moon with bronze. Halfway through the wait, so five minutes, maximum six minutes, later, I was thinking 'But silver would be nice.' And just before they came to tell us I thought 'I am going to be a little bit disappointed now with bronze.' When they showed us the result, they simply came out and opened a piece of paper and we saw we had won silver by eight hundredths of a second. Eight hundredths of a second.

That race and the medal had an impact, not only on each of us, but also on the sport. When we won that medal it turned people's mind-set because people stopped merely hoping, people now knew. People knew an Olympic medal success was possible for women whom we trained side-by-side with in the gym and by coaches we worked with and by the support set-up we had. And everyone believed that they could now do it themselves.

From Athens 2004 to Beijing 2008

In the four years between Sydney and Athens, the development continued apace. The English Institute of Sport was created, and we suddenly had the medical backup and scientific support we had never had before. We had physiotherapists and doctors on site. We had physiologists who monitored our physical data day in, day out. We had funding for more use of nutritionists. Everything started getting even more professional. We also trained with a different mind-set. We trained with the mind-set that said 'We are now here to win medals.' We took a smaller team to the Athens Olympics. There were only eight female rowers, three boat classes: a quad, a double, and a pair; and everyone went there absolutely genuinely believing that, if they could get it right, they could get an Olympic medal.

Sydney was the first medal ever won in Olympic history and then, four years later, in Athens every single woman came back with an Olympic medal for women's rowing. All eight won medals: an unbelievable, incredible turnaround. The only slight problem, if you could call it that, was the fact it was two silvers and a bronze. Suddenly the colour had started mattering. When you are operating at a high level and enjoying sustained success, then it is the gold medal everyone is after.

After Athens, people started talking about what the margins are between those final steps: to the bronze, bronze to silver, silver to gold. The talk was about the differences you can make and the marginal gains you can make. The idea is that you can make a difference – it does not matter how big or small – even a tiny difference in one element of your training. When you start adding that together between maybe two people in a boat, four people in a boat, eight people in a boat, then it starts to become a big gain. If you multiply that by all the different areas you could do it in – could you do it in the gym work, could you do it on the rowing machine, could you do it in the boat technically, could you do it psychologically, could you do it with nutrition, could you do it with rest and recovery, could you do it with all these different areas, could you make marginal improvement in all those, and of course everyone can, then you start adding that together and you start making definite steps forward. This addition of marginal gains became a major theme after Athens. The British Olympic Association, which takes us to each Olympics, made a video stressing the idea of all those tiny, minute improvements making a difference because that is how competitions are won and lost. It was a powerful way to capture exactly the margins in sport and why every single session does matter and why everything you do makes a difference.

We female rowers, therefore, came together as a squad after Athens, absolutely acknowledging our success but all agreed that to really shake things up we had to now go for gold. We felt we had the right people and the right set-up, and things had progressed enough in our sport, so that was a realistic aim now.

We were a big squad and we knew realistically not everyone in the squad was going to be the one to win the Olympic gold, but everyone could still play a key role. You need a bigger squad than the actual people who will go on and win, but everyone needs to be signed up for that mission. We still had the heart and soul and passion that Sydney had delivered, we had the utter belief that Athens had created, and then the run up to Beijing was about basically leaving no stone unturned, making everything matter, making everything count.

We worked with psychologists quite a lot. We talked far more about rules within the boat on a different level of understanding, how we can develop as athletes individually but also as a team. There is always the question of how you become a successful team, what is the x-factor that makes some teams work perfectly and some just fall apart. A lot of teams work well when far away from the pressure but, when put under the pressure of something like the Olympic Games, or the World Cup Final, or anything like that, many teams then crack. If there is any weakness in a team, it will be seen under this massive pressure. Sometimes that is the first time it is seen, but by then it is too late, and the effect can be disastrous. In the run up to the Beijing Olympics we tried to be incredibly honest about what we felt about each other and how we felt about the boat, and what was important to us. We had some sessions, which were really positive, when we took only one person at a time on the team and everyone said something about them, something that inspired you about them, or what you loved about them, or what you were in awe about with them. It was all really good confidence building. We also could say the tough stuff that you do not want to hear, some of it not all great, but there was openness and honesty leading up to Beijing as we did not want any surprises.

There were three World Championships between the two Olympics, and we won each World Championship. We won in 2005, 2006, and 2007, so we were triple world champions coming into the Beijing Olympics. We had never before had such consistent success at that level in British women's rowing, and everything was then geared up for the final step: the Olympic gold. We had the perfect preparation, we had a year where everyone was fit and healthy. The selection was incredibly competitive, which is what you want. As an athlete it is very uncomfortable, but as a coach you want it to be competitive, you want it to be hard to get into that top boat.

In the Beijing Olympic year the biggest threat in many sports was going to be China as a sporting nation itself. China had put out a very early statement saying that for the first time in history they would top the medal table. China had faced some criticism over whether they should even be hosting the Olympic Games because of their human rights history. China's response had been 'We'll put on the biggest and the best show that you've ever seen and we'll prove to the world how good we are and what a show we can do. And we'll do it not only showcasing the city but also showcasing our athletes.' Perhaps not unexpectedly, therefore, in the year leading up to the Beijing Olympics we suddenly saw their athletes coming out in all different events with absolutely exceptional performances. So, in 2008, even as reigning world champions, we knew it was not going to be simple, and it never is. Past results are no guarantee of future success.

We raced a World Cup season before the Olympics and we won some of the races, but the Chinese crew beat us in other races. Coming into the Beijing Olympics, we were equal favourites. We were confident in what we could do but we knew we would have to bring our absolute best to that event. We raced our heat confidently. We were in separate sides of the draw from China, so we did not meet them at that point. We broke the Olympic record, and then the Chinese raced their heat and beat it by more. In the final we knew, based on the times and the performances, realistically it was going to be a two-boat final: us and the Chinese.

The tactics by which we had won the year before had relied on a very fast start, then we got ahead of everyone and kept going until there was just too much of a gap for anyone to attack successfully. We decided to race the same style because it had been our most successful year, so we wanted to race in a similar way.

There is a theory about development cycles, mainly derived from business, but it is found in evolution as well: there are different phases so that everything eventually occurs in a circular way. For instance, you have your learning phase, which usually comes off the back of maybe a disappointment or something that's happened; you have to learn to get better, you are trying out different things, and you are improving the whole time. Then you have the growth phase where you have now learnt it and then you develop it, you bed it in, you get confident, and the praise starts coming. But then you have the decline phase where usually things start not working any more and you will have to begin again. The key thing, it is argued, to avoid the decline, is that you must move onto another cycle again while you are still in the growth phase. That way, you would actually change the circular nature and start learning and growing, and learning and growing, and you never hit the decline.

Therefore, with hindsight, having done our learning through the early Olympics, we did our growth phase probably in the 2005, 2006, 2007 time, and at that point we should have changed again and tried different tactics. However, we stuck with the same successful one. The risk with that was that, although it was the fastest way and we knew how to race that way, it also was known by others.

In the Olympic final we had a very fast start, we came out ahead of everyone, we pushed away, created a nice gap, and it was going exactly according to plan. A 2,000 metre race: the first 1,000 we had a good margin, the third 500 everyone starts closing up, and in the last 500 metres you get into those amazing crowds, deafening roars: the

Chinese crowds were sensational, drums pounding and everything was happening. The Chinese crew at that point started a sprint to the line that we had never seen. We had never seen them do it and did not know they could do it. Obviously we responded. We knew the threat was going to come at some point and it came then. We responded, but we could not match their speed. They lifted their speed again and we tried to respond again. At this point the line was coming closer and closer, and they were drawing level with us and we were nip and tuck the whole way. The crowd was deafening around us, we were still trying everything we could do. They started pulling away, we were about to run out of lake and we crossed the line a second behind the Chinese crew.

It was absolutely devastating. We were all crushed in an instant. All our hopes and dreams shattered absolutely instantaneously. No second chance. Everything we had worked for for four years felt like it had come to nothing. We were the crew flying the flag for the whole team, but we did not get it right in the one race that mattered. I remember we rowed to the landing stage to meet John Inverdale and Steve Redgrave, who were working for the BBC. I clearly remember their look of horror as they watched four broken women in tears coming towards them for interview. No one knew what to say, and no one wanted to speak, and no one knew what to do. We then moved to the podium for the medal ceremony, standing next to an understandably ecstatic Chinese crew in front of their home crowd.

Increasingly, one of the hardest things in the aftermath of a disappointing result is the feeling of a need to apologise for it. I know I felt that we had let people down, I felt I had let my teammates down, let our coach down, all these incredible support staff who worked with us day in, day out. You know you are representing the country, you know you are representing Great Britain, you have let down the wider Team GB, you have let down the country. You genuinely feel indebted to everyone who has ever done anything to try to help you – and I had many friends and family who had flown out to Beijing at huge expense – and I just felt I had let everyone down. You lose perspective and simply see it as a huge failure. It is quite traumatic and nothing anyone can say fixes it. No one, not a single person, ever said 'Well, you know, that was such a disappointment,' but it's all you feel.

Later, we attended a dinner and there was a woman there who had done, in preparation for the dinner, a thing on a website about 'I'm going to meet some of the Olympic athletes, if you've got any questions for any of them let me know and I'll put them forward and we'll get some conversations going online or whatever.' All anonymous, lovely. The first question I got given, on a piece of paper, was 'Can you ask the women's quad what it feels like to have wasted four years of their lives?' The next one was along the lines 'Can you ask the women's quad how they dared to show emotion on the podium?'

It shows you how far the sport developed even in four years because I really do not believe any of those thoughts would have been raised after London 2012, when you saw so many people upset over medals they were in some way disappointed with. However, at that time in 2008, for some reason people just felt it all right to raise such questions. I do not know what makes people ask these things. But it was such a sensitive time, I remember reading it and then thinking 'Oh God, is that what I've done? Have I wasted four years of my life getting that silver medal?'

The London 2012 Olympics

I took a bit of time off to make the decision whether or not I wanted to continue on to the London Olympics. I really did agonise over it, but I decided to go on – which was obviously the right decision – but at the time it felt like a hugely difficult decision. Those cruel questions had their use as I remember when, a few months later, I finally knew I was okay because I found that piece of paper again and rather than being upset and hurt by it, I was actually really very angry about it and defiant. I thought 'Do you know, I do not know who wrote that or what they have done in their four years, but in my four years I have been a world champion three times and an Olympic silver medallist and I do not think that is a waste of my life.' I think at that point I knew I was going to be all right because you know when that fighting spirit comes back. It was now game on.

The run up to the London 2012 Olympics shows how we developed, not just as athletes, but as a sporting nation, and how the Olympics have developed. London is the only city to have hosted the modern Olympic

Games three times, in 1908, 1948, and then 2012. In 1908, 22 nations competed, and there were 2,008 athletes, 37 of them were women. Thirty-seven women athletes out of 2,008 back in 1908! Then 1948 was the first Olympics since the Second World War, so they were an austerity Olympics. Britain pretty much won everything. But 2012 was always going to be different. Sport in those 50–60 years had utterly developed in a way no one had predicted. As an Olympic movement, rather than having 2,000 athletes, we had 10,700 athletes in London, almost half of whom were women. Women's sport had developed so much that, with women's boxing being introduced, the London Olympics was the first Games at which there would be a women's representative in every single sport and, because Saudi Arabia, Qatar, and Brunei had entered female athletes, it was the first Olympics to which every single nation sent a female competitor as well. A wonderful record! Therefore, women's sport had moved on, the world of sport had moved on, and, as British athletes, there was this whole different expectation of success. People expected gold medals every day.

I was in a double with Anna Watkins – an ex-Cambridge girl – and we had been together for three years at that point and we had won every race we had ever done together. So there was this huge expectation centred on the Olympic gold. Obviously I knew it was not that simple, but we were just a fantastic combination and we loved what we did. We simply loved it. We loved training every single day. Our boat was just a gift: it was fast, it was easy, it was comfortable, we had amazing communication. But we still wanted to make sure. We wanted there to be no chance that anyone could beat us when that day came. We wanted to make sure, but there are no guarantees in sport, as I said before. You can, however, eliminate as much risk as you can.

Our sport had developed so much that we had even more amazing experts in the wings. We now had sports science and sports medicine backup, we were in the shape of our lives, we asked all our wide support team what they could bring to us. We wanted them to be part of this gold medal attempt, and we asked them all to bring in their gold medal standards to see whether we could raise everything we did. It was a team effort in every way, and together we looked at it all to ensure it would be the top, highest quality. In the three years we had together in that boat, we probably spread the net widest at the beginning of those three years and looked at new projects and new designs, and at new things we could do with the boat, and developed in that way.

As it came closer and closer to the Olympic Games itself, we narrowed that down and did not want any extra distractions. Therefore, we cut away anything that was not absolutely essential. It was probably the clearest and simplest crew I have ever been in. It was very obvious what we had to do, both individually and together. We knew how to race and we knew how we would deliver that race. The biggest reason we had to make it that simple is because, when an Olympic final comes around, it does not matter how many times you have done it, it is a pretty enormous moment in your life. It is quite terrifying and you cannot control your thoughts particularly well.

There was one thing we had not predicted: we had no idea of the gigantic scale of public support that we would get. It started with the Opening Ceremony. The Opening Ceremony is a big deal. Everyone remembers the Beijing Olympic Opening Ceremony in 2008: drummers in perfect synchronicity. Eight is the lucky number in China, there were 2008 drummers drumming at 8:08, on the eighth day of the eighth month in 2008. There were some nerves when the London ceremony started over how it would compare with Beijing. Anyway, the Opening Ceremony started, and I still remember I was watching this green and pleasant field. It was all lovely and there were milkmaids running around and everything was very gentle and pleasant. Then somehow, in front of your eyes, it transformed into these vast industrial towers and people were welding things and it all changed and they were building, they were building something and it is glowing and they have made a ring and you go 'Wow, they have made a ring, that's amazing' and then it is going up in the air, and they have made five rings and, ah, it's the Olympic rings! And by then, as an athlete, you are just gone. Then we had Mary Poppins, and we had the NHS, and we had the world wide web and some of the best music, and we had ... well we had Mr Bean and then when James Bond meets the Queen. And the way they filmed it, the Queen was filmed from behind initially and everyone watching went 'Ooh do you think it is Helen Mirren or do you think it is Judi Dench?' And then she turns round and it is the Queen. And then she jumps out of a helicopter.

And then you just think 'This is going to be good; this is going to be very good.' Then overnight, every single paper the next day talked about 'the greatest show on Earth' and it had begun.

Then it became the athletes' turn and the nation wants medals, preferably gold. The next day there was Mark Cavendish in the road race, a guaranteed gold medal. Mark Cavendish, one of our best sprint cyclists ever seen. He had an incredible team of five, four of whom had won stages in the Tour de France two weeks earlier. Undoubtedly the favourites for gold – until they came 22nd. And suddenly people realised this was not going to be easy: that we can put on a great show, and we can be the home nation, and we can have all the support in the world, but gold medals are massively hard to come by, because you become an even bigger target on your home ground. The first gold medal in fact did not come until the Wednesday of the first week, but then it was the gold rush.

For Anna and me, the crowds on the morning of our final were the strongest memory of the Olympics. The crowds were unlike anything I have ever witnessed, or ever seen, or ever heard, or ever will again. One of the things I get asked most about London is 'Those crowds, could you actually hear them when you were racing because obviously with that tunnel vision and concentration, you're really focused?' What I would love to do is simply put anyone into that cauldron. You do not just hear it, you actually feel that noise as a physical reaction.

There are things that happened in that final that will never happen again in my rowing career. At the start line you are 2,000 metres away from the crowd. There were six lanes, and they called each nation one by one. When they announced our names for Great Britain, we heard a noise and we did not know what it was. Later we found out that was the crowd at the finish line cheering. So 2,000 metres away we heard the roar of the crowd.

In Olympic racing they block 100 metres from the start so that no one can get into that first 100 metres. So it is pin-drop quiet. It is terrifying and all you hear is your heart pumping and you are hoping the opposition does not hear it. We were focused, thinking keep it simple, keep it clear, we know what we are doing.

The race started: the first 100 metres exactly as we wanted; we are locked in what we need to do. 100 metres, we start to hear people on the bank because they have opened up virtually the full length of the course, so some people have walked all the way towards the start, so we start hearing voices then. We get to 500 metres gone, we are starting to pull out, we are leading and we can hear more and more people on the bank. By 1,000 metres there's so many people that it is like a normal finish line for us. It is really very loud.

750 metres to go. Thankfully we are still leading. It really starts to become a deafening noise. We cannot hear what is going on in our boat or communicate with each other.

500 metres to go, and you then hit the actual stadium. We had not hit the stands yet, these are just people on the bank so far. We then hit where the stands are and the noise goes insane. They were metal stands, people were stamping their feet and roaring and the whole thing – which became known as the 'Dorney Roar' – was just crazy. Because the stands are built obviously either side of the water, the noise drops down on the water and it gets magnified by the water. So this sound is getting louder and louder. It had an effect that I have never seen before: it was causing vibrations in the water, so the boat was actually reacting to the crowd.

Anna, who sits close behind me, is shouting at the top of her voice, I have no idea what she is saying, I think 'go' was probably it, I am pretty sure that's what she was doing. If you watch the video of the race, you see how she gives a little bit of a smile because she knows we are going to win. I still look like I am going to kill someone because I tend to feel you never know for certain until you cross the line.

Then, when we approached the line, the sound was getting louder and louder and louder. I thought my head was going to explode. As we crossed the line the crowd just could not have gone any louder. It was already at maximum volume, but what changed was the tone of it: it went from this 'come on!' roar to a massive celebrative noise when we crossed first and it flashed up 'Great Britain, Olympic Champions'.

It genuinely was the most amazing moment of my life. And then, what was amazing was how it was enhanced every single day after that. We kept meeting people who had been inspired and moved and touched and enriched by the whole Olympic Team, and we realised we were part of something very special. If you choose sport as your career, some people think it is quite a selfish thing and it is quite a single-minded obsessive thing to do. What I came to realise was how it can broaden, reach, and touch so many, and enrich people's lives.

I think what we did as a nation was phenomenal, and we helped to make sport accessible and it reinspired people. The idea always was to inspire a generation, and by that everyone meant the new generation, the young generation. However, I was stopped by a man in his early eighties who said to me 'Dear, you have not just inspired the new generation, you have inspired every generation in this country.' And you just think something very special happened that summer, and hope that this spirit will live on beyond all of us.

Concluding Reflections

To have been a little part of that is wonderful. I went to university to become a big shoulder-padded lawyer but I left trying to make a boat go backwards. It was not the development I thought would happen in my life. It has taught me amazing things about what can happen if you have a vision, and if you have a dream and you have a belief in it, and you pull together an extraordinary group of people who will take things on and rise to those challenges, and learn from mistakes, and grab the opportunities they get.

One of the big things for me is that development is not guaranteed or constant. Sometimes it is painful, and you need to go through disappointment and failure to really move onto the next level, and it all really taught me not to be afraid of that failure because it is just a massive learning process. Then sometimes you get an opportunity during that development phase to really make a leap and to make a stand, and you have got to grab those moments.

I think everyone else, not just me, thought that this was the end point; that it was a natural end point in my life and my career. I took two years to calm down from London to be really honest. Then, in 2014, I had to make a momentous decision – and I made the decision to get back in a boat. Now, at the time of writing these lines, I am back training with the Rio Olympics in mind. A lot of people have understood why and a lot of people have not. A couple of national stories in the press said it was a huge mistake trying to go back because when you have got something that is so perfect, a crew that is so perfect, and an end point that is so obvious, I could never top it, I could never beat it, why would I ever try and go beyond it.

All I can say in answer is to echo something David Brailsford, the cycling Performance Director, said recently. He said sport is not about an end point and about a result. It is a journey, and it is a journey of development. It is about learning more about yourself and about what you can do and what you can do with other people. It is up to you at what point you want to end it. It is not for other people to say, or because of an age thing, or because of anything. It is an ongoing journey.

For me, it has been the most rewarding, challenging, satisfying, enriching journey possible. And I have got a chance to continue on that journey a little bit longer. I firmly believe it is possible to do things that I never thought I would before, and it is something I love more than anything else in the world, so why would I not do it? I am not going to be told by other people who have never stood in my shoes that it is the wrong thing to do.

Of course it is a risk. I might be in tears again in two more years' time, who knows, but if you are not willing to stand up and take a chance on what you do and find out more about yourself then I just think, what is life for? We have got one chance at it, and you have got to make the most of what you have got while you are here and do what you love, and love what you do, and hope to make a difference in some way.

I get to do that in a boat, and I am going to keep doing it as long as I can and as successfully as I can. If all goes well I will come back and tell you about the next stage in my development.

3 The Development of Galaxies

RICHARD ELLIS

Time Travel and Development

It is a familiar dream that we might one day be able to travel back in time. Imagine going back to interview William Shakespeare, or being present to witness in person the construction of the great pyramids. Observational astronomers like myself are unique as scientists in that we can witness the past. As the speed of light is finite, when we look at the Sun we are seeing it as it was just over 8 minutes ago; that is the 'light travel time' to the Earth. The nearest star is seen as it was over 4 years ago, and light takes 70,000 years to traverse the Milky Way, our own galaxy.

The situation is transformed when we take an exposure with a powerful telescope into deep space. Figure 3.1 is the deepest near-infrared picture taken with the Hubble Space Telescope; it represents a total exposure of nearly 10 days' duration of a small patch of sky called the 'Ultra Deep Field'. In angular extent, it is about a tenth the diameter of the full moon. The most distant galaxies marked in this image are being seen as they were, not thousands or even millions of years ago, but 13.5 *billion* years ago – a time well before the Earth and Sun were formed. As we believe the Universe itself is 13.8 billion years old, we are therefore witnessing young galaxies in formation when the Universe was barely 3% of its present age. In a way, we are looking directly at our origins.

If we can estimate reliable distances to remote galaxies such as these, then this opens up the prospect of ordering them in 'light travel time'. Such 'time-slices' can then be joined together to witness how galaxies develop and evolve. The trick was familiar to us in childhood: we drew a developing scene on the successive blank pages of a notebook and then

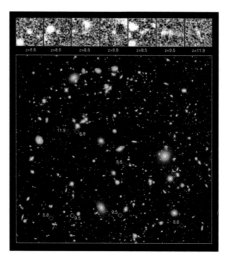

FIGURE 3.1 The deepest near-infrared image of the sky to date taken with the orbiting Hubble Space Telescope. The image consists of exposures through various colour filters totalling over a week of collective exposure time. The area viewed, called the 'Hubble Ultra Deep Field', covers a diameter about a tenth that of the full moon. The coloured boxes represent the locations of the seven most distant galaxies that have been located in this image; their likely redshifts are also indicated (z-values). Individual images of these galaxies are shown at the top, from left to right in order of distance and hence 'look-back' time. The most distant are being seen over a light travel time of 13.5 billion years, corresponding to 97% of the way back to the birth of the Universe. (Original images courtesy of NASA.)

flipped those pages rapidly in sequence to create a primitive movie. Similarly, we can use large telescopes to directly witness the evolution of the Universe.

I will discuss two forms of 'development' in this essay. The first is the 'development of galaxies'; understanding the 'movie' made possible by the time travel mentioned above. The challenge is not only to connect the various time-slices, but to develop a physical understanding of the evolution we observe. The second form of 'development' is the use of new technologies or innovative ideas to accelerate progress, that is the development of our observational capabilities. As we will see, progress in observational astronomy is typically governed by new or more powerful facilities enabled by technological breakthroughs.

The Morphology and Structure of Galaxies

Before we attempt to create an evolutionary movie, let us consider what we mean by a galaxy and why tracing its evolution is so important. The Milky Way is familiar to anyone who can witness a moonless sky away from city lights. Perhaps surprisingly, its physical nature was not properly identified until Thomas Wright of Durham (1711–1786), a landscape architect, conjectured in 1750 that the band of light across the sky represents a thin structure composed of myriads of unresolved stars. Unfortunately, although Wright got the correct idea (one that influenced subsequent natural philosophers including Immanuel Kant), he embellished it in his publication *An Original Theory of the Universe* with more fanciful ideas, including the existence of a spherical shell in which the Sun is embedded and some form of superior being residing at the centre of that shell.

Until the 1920s, the Milky Way was considered by many astronomers to be the entire Universe. The Sun was inferred to lie in its outskirts because the Milky Way is more spectacular when viewed from the southern hemisphere. Although some astronomers, including William Herschel (1738–1822), speculated that there might be other 'island universes' beyond the Milky Way, to demonstrate these are external systems required knowing their distances. This was first achieved by Edwin Hubble in 1923 via the demonstration that the Andromeda spiral, Messier 31, lay at a distance well beyond the confines of the Milky Way. The Andromeda spiral, our closest comparably sized neighbour, is 2.5 million light years away. The term 'galaxy' became commonplace from that time onwards, and has its origin in the Greek γαλαξίας, meaning 'milky one'. Hubble demonstrated that the Universe was far more vast than had previously been imagined.

Galaxies are stellar cities containing as many as 10 thousand million stars of various colours, masses, and ages. While many people are probably familiar with their varied morphological forms – elegant *spirals* with central bulges, smoother *ellipticals* devoid of such features, and *irregulars* which have no clear symmetry, the astronomer is also concerned with their three-dimensional and dynamical properties. The question of how we can determine their three-dimensional forms therefore arises.

I am reminded at this point of a public exhibit being planned at the Royal Observatory Edinburgh, where a staff member was charged with constructing a plasticine model of a galaxy based on a colour photograph in a popular astronomy book. Unclear about its three-dimensional nature, he went to an astronomer at the Observatory and enquired 'Do you have a shot of this thing taken from a different angle?'

One can certainly piece together some idea of the three-dimensional shape of certain classes of galaxies by assuming they represent similar systems viewed at random angles. However, a more robust approach involves understanding the internal motions of the constituent stars. If we pass the light from different portions of a galaxy through a spectrograph, then, using the Doppler effect, we can deduce that spirals are thin disc-like structures rotating at velocities of around 200 km/s such that a complete rotation takes about 200 million years or so. This is much less than their ages, so you can see that their disc-like shape directly reflects their rotation. Elliptical galaxies, on the other hand, are found to hardly rotate at all. We believe their three-dimensional shape reflects the envelope of the random orbits of their constituent stars. Perhaps surprisingly, these 'ellipticals' are not flattened by their rotation.

There are further profound differences between these two important galaxy types. Spirals contain copious amounts of hydrogen gas, detectable using radio telescopes, and are actively forming young blue stars in their spiral arms. In contrast, ellipticals generally have little gas and contain mostly old red stars. As hydrogen gas is the fuel from which young stars form, ellipticals are sometimes referred to as 'passive' or 'quiescent' systems, whereas spirals are continuing to form new generations of stars.

One final remark about spirals and ellipticals from the perspective of the observing astronomer. Although spirals are more numerous in the local Universe, it is indeed fortunate for us that we live in one. Had we been born in an elliptical galaxy, the night sky would be ablaze with red stars in all directions and we would have a very hard time seeing any external galaxies, at least with optical telescopes. So it might have taken us a lot longer to deduce the enormity of the Universe.

Looking Back and Redshift

The key to measuring distances in space and hence being able to 'look back in time', which will enable us to make our evolutionary movie, is the *redshift*. This is the shift in wavelength of spectroscopic features due to the cosmic expansion. It is a popular misconception to imagine the expansion of the Universe as the hurtling of galaxies as projectiles into empty pre-existing space from a particular point, say the centre of the Universe where the Big Bang itself happened long ago. This is an incorrect interpretation. In fact, on average galaxies are not really moving at all. It is the space in between galaxies that is being stretched without the implication of any cosmic centre. The redshift measures the effect of this stretching on the wavelength of light received by the observer. It therefore encapsulates how much the Universe has expanded since the light left its source (Figure 3.2). If we can measure the redshift using a spectrograph, comparing the wavelength of key diagnostic features with their emitted values as deduced in a laboratory, we deduce how far back in time we are probing – the *look-back time*. A redshift of zero corresponds to a local galaxy seen at the present epoch; the Universe has not expanded significantly during the light travel time. A redshift of 1 corresponds to a doubling of the wavelength of a particular feature (the relevant shift is 1 plus the redshift), and this means looking back about 7 billion years in cosmic time – just over halfway back to the beginning of the Universe.

To take a picture of a patch of sky and find the positions and brightnesses of a set of galaxies is relatively straightforward with a large telescope, but to secure the spectrum of one and measure its redshift (or look-back time) requires much longer exposure times, given that the light must be split by a prism or diffraction grating into its constituent colours. The first technological development I want to discuss is called *multi-object spectroscopy*; it has made this task much easier and began while I was at Durham University in the 1980s. The focal plane of a telescope offers a certain field of view on the sky and, just as we can take a picture of that field, so, using optical fibres, we can feed the light from a set of galaxies within that field of view into a spectrograph – a form of optical plumbing. Initially we did this manually, plugging fibres into a brass plate with

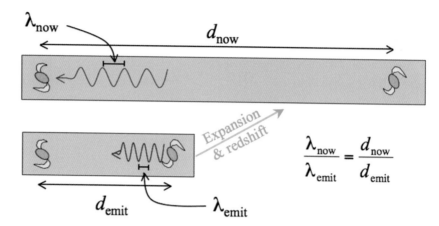

FIGURE 3.2 The concept of 'look-back time' and redshift. As the Universe is so vast, during the time interval between the light ray leaving one galaxy and arriving at another (say our own Milky Way), the intervening space has expanded a significant amount. This 'stretching of space' is the correct interpretation of the cosmic expansion and increases the wavelength of the light rays. In the lower panel, a galaxy on the right emits light of wavelength λ_{emit}. The receiving galaxy on the left is a distance d_{emit} away. By the time the light arrives, the Universe has expanded, the separation is now d_{now}, and the wavelength has been stretched to λ_{now}. The ratio of the two wavelengths, which is (1 + the redshift), is in fact the ratio of the two distances $d_{\text{now}}/d_{\text{emit}}$ and hence a measure both of how much the Universe has expanded in the meantime and of the 'look-back time' to the galaxy being viewed. (Courtesy of Mark Whittle, University of Virginia.)

pre-drilled holes at the precise locations of 50 faint galaxies. Later we automated the motion of these fibres using a robotic arm equipped with an electromagnet that picked each fibre and placed it appropriately. To illustrate the versatility of a robotic positioning system, at the time we used our software to configure the fibres in the pattern of a map of Australia (Figure 3.3(a)), to which a frequent audience response was 'What about Tasmania?'

Via more ambitious positioning schemes, we successively went from gathering redshifts 100 at a time to 400 at a time. This culminated in the so-called '2 degree field redshift survey' whereby a team of UK and Australian astronomers surveyed 250,000 galaxies, making the first systematic map of the three-dimensional structure of the Universe out to a redshift of about 0.3, corresponding to 3–4 billion years ago

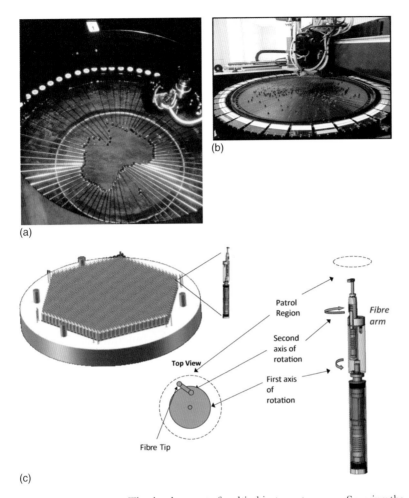

(a)

(b)

(c)

FIGURE 3.3 The development of multi-object spectroscopy. Securing the spectrum of a galaxy is key to understanding the time in cosmic history when it is being viewed, but this is costly in telescope time. Gathering many spectra simultaneously is therefore highly advantageous. Optical fibres, which can convey the light of many galaxies in the field of view of a large telescope to a spectrograph, can enable this, provided a means is found to position them accurately on their targets. During the 1980s a robotic positioning technology was developed at Durham, and this has led, in its most recent incarnation, to a multi-object spectrograph being built for the Japanese 8 metre Subaru telescope capable of surveying 2,400 galaxies simultaneously. (a) The Autofib robotic positioner at the Anglo-Australian Telescope (100 fibres, c. 1985) configured as a map of Australia. (b) The 2 degree field positioner (400 fibres, c. 1998). (c) The Subaru Prime Focus Spectrograph under construction (2,400 fibres). The Subaru system has independently controllable units (see detail of one of these) using two piezo-electric motors which drive a fibre-optic cable at its tip to any point within a small patrol region. All 2,400 fibres can be accurately positioned independently and simultaneously in only 40 seconds.

(Figure 3.3(b)). Most recently, at Caltech we are building an instrument for the Japanese 8 metre reflecting Subaru telescope in Hawaii that will gather the spectra of 2,400 faint galaxies simultaneously (Figure 3.3(c)). This instrument will survey up to 10 million galaxies out to early cosmic times and measure how the large-scale structure of the Universe has grown over more than half its history.

At Durham, our research group was particularly interested in studying the properties of very faint galaxies, so we used these instruments and others eventually to reach to a redshift of 1, corresponding to a look-back time about 7 billion years ago. Those spectroscopic surveys were statistically complete, in the sense that we chose a few patches of sky and studied every galaxy within them down to a brightness limit that could be reached in a reasonable exposure time with the 4 metre telescopes available to the UK at the time in Australia and the Canary Islands. These surveys gave us our first view of evolution in the galaxy population as I discuss below.

However, in the 1990s, when the more powerful twin Keck 10 metre telescopes came online in Hawaii, a clever technique was developed for locating distant galaxies via their colours. It is quite inefficient to select galaxies only according to their brightness if one is interested in pushing to great distances. A fainter object need not necessarily be more distant; it could be a more feeble object close by. One of my Caltech colleagues, Chuck Steidel, demonstrated that it is more efficient to first screen all galaxies using a trick exploiting the observation that below a certain ultraviolet wavelength the light of all galaxies is suppressed by hydrogen gas in the galaxy. Using three optical colour filters to locate those galaxies which disappear at the shortest wavelength enabled Caltech astronomers to find the first galaxies at a redshift of 3 – corresponding to looking back 10–11 billion years, more than three-quarters of the way back to the beginning. The same technique – colour selection – was used to find galaxies as shown in Figure 3.1 at a redshift of 10 – 97% of the way back to the beginning.

A Simple Evolutionary Picture

With these techniques in hand – spectroscopically determined redshifts, aided by both multi-object spectrographs and later colour selection, we

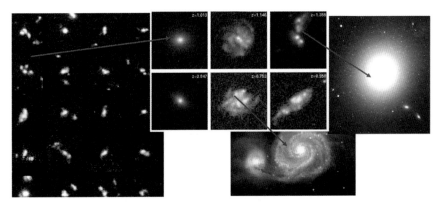

FIGURE 3.4 A time-slice of the Universe over the past 10 billion years, corresponding to the last 75% of cosmic history. Each panel represents a mosaic of galaxies viewed with the Hubble Space Telescope but selected in different redshift intervals. From left to right: redshift 3 (when the Universe was 3 billion years old, a look-back time of 10 billion years); redshift 1 (age 5 billion years, look-back time 8 billion years); and redshift zero (the present). Over more than half the age of the Universe, galaxies have grown in size and become more regular in form. The red arrows indicate the leaps forward in time from redshift 3 to redshift 1, and then onwards to 0. (Original images courtesy of NASA.)

are ready to join together our first time-slices. In Figure 3.4, I show Hubble Space Telescope pictures of galaxies with known redshifts, sorted according to three look-back times: today, 8 billion years ago, and 11 billion years ago, corresponding to redshifts of 0, 1, and 3, respectively. A number of patterns emerge.

Firstly, if we consider a redshift of 1, or 8 billion years ago when the Universe was just under half its present age, there's a sense of familiarity (shown on the panel of galaxies in the middle of Figure 3.4). We see galaxies that resemble present-day ellipticals and similarly there are some spirals, although they appear more ragged and perhaps less mature than those we see around us today. We also find many irregular-shaped galaxies; although irregulars are seen today, they are relatively rare. A natural deduction is that the basic morphologies of galaxies were already in place at this time and that subsequent evolution has largely been minor, with the possible exception of a declining abundance of irregular galaxies.

However, if we leap further back to a redshift of 3 (shown in the panel on the left), two billion years earlier, we cannot locate any galaxies that resemble present-day systems (shown on the right). Most are highly irregular in shape, often with multiple components as if assembling from smaller components, and, above all else, they are physically much smaller. A typical galaxy at a redshift 3 is only 6,000 light years across, compared with our Milky Way, which is 70,000 light years in diameter. Clearly this period, when the Universe was 2–5 billion years old, was a formative one in the development of galaxies.

Since I moved to California, my colleagues and I have focused on this period of cosmic history. We attempted to understand the internal motions of the gas and stars in typical galaxies at that epoch since, as we saw earlier, similar studies were pivotal in giving us a clear picture of the three-dimensional nature of local systems. This is exceedingly challenging work, however, as such distant galaxies are exceedingly faint. Although it is relatively straightforward to take a picture of such a remote galaxy with the Hubble Space Telescope, it is much harder to dissect the light from each part of that galaxy with a spectrograph so as to deduce the relevant internal motions. This requires a telescope aperture much larger than that of the Hubble Space Telescope (2.5 metres), and hence we turn to a ground-based giant like the Keck 10 metre telescope.

Disappointingly, a distant galaxy is simply a blurred point of light when viewed from the ground because of the turbulent layers of the Earth's atmosphere. In the last decade, enormous strides have been made in my second technological development: *adaptive optics* – a technique for correcting for this blurring, rendering an image sharper even than that secured with the Hubble Space Telescope. The idea is simple: we shine a 10 watt laser into the sky, and this illuminates a layer of sodium atoms 80 km above. The spot on the sky illuminated by the laser represents an artificial star close to our target and we monitor its incoming signal with a sensor that is sensitive to the imperfections induced by the Earth's atmosphere. This sensor drives a deformable mirror in real time that corrects the distorted signal from our target. People often find it hard to believe that such a technology, using a ground-based large telescope, can produce images even sharper than

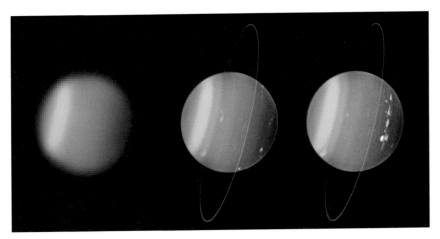

FIGURE 3.5 Adaptive optics with a ground-based telescope can deliver images sharper than those taken with the Hubble Space Telescope; a dramatic illustration with the planet Uranus. The left image is an infrared image without adaptive-optics correction, and the right two images (courtesy of Lawrence Sromovsky, University of Wisconsin–Madison/W. M. Keck Observatory) illustrate the gain in resolution achieved by adaptive optics, enabling clouds of methane to be seen changing over a period of 24 hours.

those taken with the Hubble. However, it can, and the results are striking, as demonstrated by observations of the planet Uranus with and without adaptive optics (Figure 3.5).

Such high-quality imaging exploited by using adaptive optics confirms that even these clumpy irregular small galaxies at a redshift of 3 are rotating. They are not rotating as fast as the Milky Way, but they are rotating nonetheless. The clumpy structures represent unstable regions of a primitive disc. Dynamical considerations suggest that these clumps will gradually migrate to the centre of the galaxy, forming a bulge. Although the spiral arms have not yet developed, we are witnessing the earliest stages of such disc galaxies. Likewise, we also find that early ellipticals contain old stars. They are quite compact and dense and seem to be the nuclear seeds of the larger, more diffuse, ellipticals we see today. So, in both cases, there appears to be some form of inside-out growth whereby early systems accrete material, perhaps from captured small galaxies or inflowing gas from the space in between galaxies.

Modes of Galaxy Assembly

As a result of this observational progress, attention is now focusing on what physical processes are responsible for this size growth of developing galaxies. As we have seen at a redshift 3 when the Universe was only 2–3 billion years old, all galaxies were much smaller than they are today. But many contain lots of gas and are forming stars prodigiously. A typical galaxy might be 10 times smaller than the Milky Way, but it is forming stars 20 times faster. Somehow this phenomenally active period subsides. The galaxies at redshift 3 are being seen at a key moment in their history; in some sense they are the energetic adolescents, whereas the older galaxies viewed today are slowing running out of gas.

Hydrogen gas is the natural fuel for young stars, and it is plentiful in the regions around galaxies, but this gas must be drawn into a galaxy and collapse into compact regions where new stars can form. To clump on the relevant scales it must first cool down. Any mechanism that keeps the gas hot will therefore inhibit star formation. Paradoxically, a burst of star formation can heat up the residual gas, preventing it from forming further stars. We call this kind of behaviour a 'feedback loop' – gas is available to form stars but some form of energy injection prevents this from happening. Two types of feedback are particularly effective. Hot stars quickly explode at the ends of their short lives via *supernovae* – these supernovae deliver enormous amounts of energy into the gas and, in smaller galaxies, may expel the gas entirely out of the system. Material falling into a *massive spinning black hole* in the centre of the galaxy can also lead to high-speed jets that do the same trick. We know that supernovae are more numerous and that black holes are more 'active' at early times, so they are probably the dominant sources of early feedback.

The fate of a young galaxy thus depends on the competition between growth fuelled by infalling cold gas, or perhaps from the merging of galaxies, and mass loss from any outflowing gas. The key to distinguishing these competing elements is the *chemical composition of the gas.* Infalling gas from outside the galaxy will be pristine in terms of its chemical make up, comprising pure hydrogen and helium unpolluted by the nuclear burning that occurs only in stars which produce heavier elements like oxygen, nitrogen, and iron. However, any outflowing gas, driven for example by supernova explosions, will have been enriched by stellar processing and hence will be rich in these heavy nuclei.

Locally, we find isolated galaxies have *chemical gradients* across them. The nuclear regions are richer in heavy elements than their peripheries. Conversely, merging galaxies show no chemical gradients; the gas from different parts in each galaxy has been thoroughly mixed in the merger process, rather like ingredients in a cooking bowl. Observational astronomers have realised that the key to understanding feedback is to trace these chemical gradients over cosmic time as well as in systems of different morphologies and dynamical structure.

Here a third development, *gravitational lensing*, can lend a hand. A remarkable feature of Einstein's successful theory of gravity is that light rays travelling across the Universe are bent by space that is shaped by massive objects. As was first verified by Arthur Eddington in 1919 at the time of a solar eclipse, if we steer our telescopes in the direction of a dense cluster of galaxies, occasionally we see the light of a much more distant (and hence higher-redshift) source highly distorted and magnified by the foreground cluster. As with a magnifying glass, the foreground cluster (the lens), enlarges the angular size of the distant galaxy (Figure 3.6). Provided we understand how the optics of this remarkable phenomenon works, we can secure exquisite data on the chemical gradients of distant galaxies.

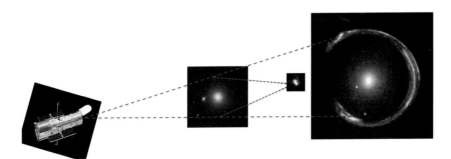

FIGURE 3.6 Magnifying distant galaxies using gravitational lensing – a remarkable phenomenon whereby light rays can be bent by massive objects. In the example shown, a foreground galaxy stretches the light from a distant galaxy almost into a complete ring, enlarging it and thus enabling more exquisite measures of its internal properties. The combination of adaptive optics (Figure 3.5) and gravitational lensing has enabled detailed studies of the internal motions and distribution of chemical elements in early galaxies. (Courtesy of Tommaso Treu, University of California, Los Angeles.)

This is pretty challenging work and, after several years' effort, my team has secured chemical gradients for only 20 galaxies at redshifts up to 3. But, broadly speaking, the results agree with those of numerical simulations of the assembly process which include the effect of feedback. For most isolated galaxies, those at high redshift show very steep gradients, consistent with the growth arising from material being added subsequently at the peripheries. Mergers, on the other hand, routinely disrupt this elegant inside-out assembly.

First Light in the Universe and the Future

I'd like to conclude by returning to the beginning of my essay, namely the quest for the earliest galaxies. Using the colour selection technique I mentioned earlier, the Hubble Space Telescope, equipped with a powerful near-infrared camera, has located several hundred galaxies beyond a redshift of 7, corresponding to the first billion years of cosmic history. It's not just a race for the most distant object, as journalists like to report, but also a search for our origins. We believe galaxies formed initially from dark hydrogen clouds in a period called the 'Dark Ages', probably 100–200 million years after the Big Bang. These clouds of pristine gas collapsed at some point, and this led to an important phase, called 'Cosmic Dawn', when the Universe was bathed in starlight for the first time (Figure 3.7). Can we witness this last frontier directly with our telescopes?

Hubble data from this near-infrared camera has shown us that the number of star-forming galaxies we can see per unit volume is declining fast as we reach to these early times. We are getting close, therefore, to the epoch when there were no galaxies at all. One way to detect Cosmic Dawn would be to witness a sharp drop in galaxy numbers beyond which none can be found. Unfortunately, it's not obvious that Cosmic Dawn was such a sudden event; it might have been a gradual process with starlight emerging first in dense environments where there is a lot of material and later in locations of lower density.

One observation that would clearly characterise the first generation of stars would be the absence of chemical enrichment. Recall that stars are nuclear factories, converting hydrogen and helium into carbon, oxygen, and heavier elements. These products are expelled into the gas during supernova explosions, so subsequent stellar generations will contain

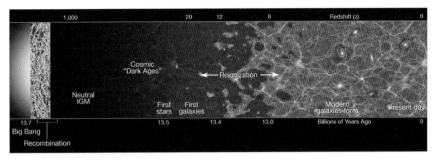

FIGURE 3.7 A timeline from the Big Bang (left) to the emergence of the first galaxies. In the first 370,000 years, the Universe was composed of a hot plasma whose radiation emerged as the cosmic microwave background when the hydrogen atom first formed. The Universe then entered the so-called 'Dark Ages' during which hydrogen clouds clumped under gravity and eventually collapsed to ignite the first stellar systems, an event popularly termed 'Cosmic Dawn'. These early galaxies contained very hot stars, unpolluted by nuclear processing, whose intense ultraviolet radiation broke apart the hydrogen atoms in space into their constituent protons and electrons, a process called 'reionisation'. These ionised bubbles became more numerous and grew in size as further galaxies were born, and eventually the entire volume of intergalactic space was fully ionised. (Courtesy of Brant Robertson, University of Arizona.)

these elements. The clearest signature of a first-generation stellar system would be the absence of these heavier elements.

I would therefore argue that the route forward lies with chemical analyses of early galaxies. Unfortunately, with our current facilities – the Hubble and Spitzer Space Telescopes, the twin 10 metre Keck telescopes and equivalent European ground-based telescopes in Chile – measuring the chemical composition of the gas in such early galaxies is too difficult. We need more powerful facilities. Fortunately, these are imminent. In 2018, NASA and the European Space Agency will launch the 6.5 metre James Webb Space Telescope, an observatory whose collecting area is seven times that of the Hubble Space Telescope but, more importantly, has an infrared capability that will permit access to chemical diagnostics in the earliest galaxies. And on the ground, Californian and European astronomers are constructing two giant telescopes – the Thirty Meter and European Extremely Large Telescopes – which will be completed in about the year 2023 (Figure 3.8). Therefore, the future is bright, as we probe this final frontier in cosmic history with these upcoming facilities.

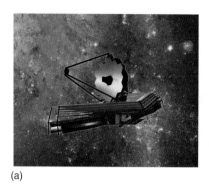

(a)   (b)

FIGURE 3.8 The promising future. Astronomers have continually strived for more powerful facilities ever since Galileo's original telescope over 400 years ago. Together with the James Webb Space Telescope ((a), artist's conception of JWST in space, courtesy of NASA), to be launched in 2018 as the Hubble Space Telescope's successor, a new generation of ground-based giant telescopes is under construction, including the Thirty Meter Telescope (TMT) on Mauna Kea, Hawaii ((b), courtesy of TMT International Observatory). The TMT's giant mirror will be made of 492 individually controlled hexagonal segments, a technical design pioneered with the twin Keck 10 metre telescopes.

In closing, let me return to the words of Edwin Hubble, whose name graces the space telescope that has been so instrumental in pushing back the frontiers I have described here. Although Hubble remains a controversial figure in the minds of some historians, he did have a wonderful talent for the dramatic. In a series of lectures given at Yale University in the 1930s, written up in his famous book *The Realm of the Nebulae* (recently reprinted by Yale University Press), he aptly summarised the adventure of pushing our large telescopes further and further into the past:

> At the last dim horizon, we search among ghostly errors of observations for landmarks that are scarcely more substantial. The search will continue. The urge is older than history. It is not satisfied and it will not be oppressed.

Acknowledgement

I would like to thank my hosts at Darwin College for the wonderful hospitality shown to me during my visit to Cambridge in January 2015, when I gave the Darwin Lecture this essay is based upon.

4 Developing a Sense of Self

BRUCE HOOD

What could be more important to each and every one of us than to have a coherent sense of self? A compelling experience that we are an integrated, unified individual occupying our bodies, the authors of our actions initiating movements, contemplating our thoughts, reviewing our histories and making plans for the future.

It is important because we know that for some people who experience disruption of their sense of self, through the effects of disease, damage, or drugs that alter the workings of the brain, life can be very difficult. Indeed, all of us can experience temporary episodes where our sense of self can be disrupted and we do things which seem to be out of character.

In such circumstances, we may say 'I was out of my mind,' 'I was not myself,' or 'It was the wine talking.' However, when you examine those excuses, clearly they are feeble. Wine does not talk. If you were out of your mind, where did you go; and if you were not yourself, then who was controlling your thoughts and actions? These aberrations reveal the true nature of self; namely that there is no integrated self as such, but rather, we are a mixture and multitude of competing urges and impulses. Some of these influences we are not aware of, and sometimes these are triggered by external events out in the world. We are not the integrated individuals that we think we are, which is why I have called it 'the self illusion' (Hood, 2012).

To say something is illusion does not mean that there is no experience. Assuming we are mentally balanced, then each of us should have the compelling experience of selfhood, and to deny that is ludicrous. Others misunderstand the premise under consideration and dismiss it offhand because all experiences require an experiencer and *ipso facto*, the self must exist. While it is tempting to enter into logical debate, it is undeniable

FIGURE 4.1 Kanizsa subjective contour illusion.

that the self is made up of experiences. Illusion in this context means that what we experience as a 'self' is not what it seems.

To illustrate the point with a visual metaphor, consider a typical Kanizsa illusion (Kanizsa, 1979), a simple example of what perception psychologists call a 'subjective contour' (Figure 4.1).

Most of us looking at this configuration perceive a white triangle sitting above the three dark circles. But what is remarkable about this triangle, of course, is that it is not there at all. If you take each circle away, the triangle disappears, so it is just an emergent property of the surrounding contours. What is most remarkable is that various brain imaging techniques that measure the activity of neurons in the visual processing areas located in the back of the brain reveal that the brain circuitry that responds to real triangles in the world also registers the hallucination that your brain creates (Von Der Heydt *et al.*, 1984). So it would seem that as far as the brain is concerned illusion and reality are not that different.

One might dismiss illusions as tricks of the mind, which is why they can be so entertaining as we delight in being fooled, but the truth is that much of our daily conscious experience is also illusory. For example, as you read this text, you probably think that you see every word in front of you accurately but you are only processing a small percentage of the visual field – everything is blurred outside of your foveal view. There are two gaping holes that correspond to the visual blindspot in each eye caused by the lack of receptors in the optic disc and yet you are not aware of them because your brain fills in the missing information. Every time you move your eyes, saccadic suppression turns off visual input so that the scene does not smear, which is why you cannot observe your own eye movements when looking in a mirror. Look in a mirror and focus your

FIGURE 4.2 Subjective brightness illusion triggers pupil reflex. (From Laeng & Endestad, 2012.)

gaze on your left eye and then your right. Alternate between the two. As you refixate, you are moving your eyes but you cannot see these saccadic movements. If you think the movements must be too subtle to notice, then observe someone else performing the same manoeuvre. Now the movements are obvious. You can see other people's saccadic movements, but not your own, because your brain shuts off this information. If you add up all the saccades and blinks you make, you are effectively blind on average for 2 hours in a typical day.

Another compelling example is the brightness illusion illustrated in Figure 4.2. The centre of the configuration on the left looks brighter than the centre on the right. In fact they are the same brightness. Like the Kanizsa example, the illusion is generated by the surrounding context. If you cover this up you can prove to yourself that the two centres have the same brightness. Is this all in the mind and imaginary? Yes and no. Our minds generate the experience, but this has tangible consequences for our bodies. There is a reflex that acts to protect the sensitive photoreceptors by constricting the pupil to limit the amount of light entering the eyes. This is why your pupils constrict on bright sunny days but dilate when entering darkened rooms. If you measure pupil dilation when looking at the patterns in Figure 4.2, even though the amount of energy entering the eyes is identical, the pupil reflex is triggered by the illusory brightness illusion

(Laeng and Endestad, 2012). Illusions may be tricks of the mind, but they have tangible consequences for how our bodies respond.

The brain does a wonderful job of integrating information from the external world to make it coherent; knitting it together seamlessly so we do not notice all the gaps. As my colleague Richard Gregory observed, the brain is a hypothesis generator – creating plausible accounts of the external world (Gregory, 1966).

Not only do we process the external world, but we also process an internal one as well. Our brains accumulate vast stores of knowledge over a lifetime and use this information to interpret new experiences, which, in turn, become assimilated into the existing representations, thereby updating and changing them. The best examples of this dynamic process are to be found in the case of memory. Contrary to popular common sense, memories are not static snapshots, but are active representations that the brain is constantly re-constructing and revising. We are not aware of this process to the extent that we can experience false memories that seem so real. For example, Elizabeth Loftus has shown that eye-witness testimony is remarkably fallible in that simply asking a leading question or suggesting that an event occurred can make witnesses who initially rejected the suggestion later incorporate the erroneous event as part of their bona fide memory, unaware that it is false (Loftus, 1975).

Our brain generates stories not only of the external world, but also of the internal. One of its most compelling characters at the centre of these narratives is the self. The brain generates this character, the self, as a summary of the complexity of the output of the multitude of systems in order to keep track. It is a great way of keeping track of the real complexity of the influences that are really generating our thoughts and behaviours.

From a pragmatic point of view, the self is also a convenient summary that enables interaction. We interact with selves and not complex configurations of influences and experiences through the notion of a 'self concept'. The field of social perception has demonstrated that, when we initially interact with others, we rapidly and automatically categorise them in order to adopt the appropriate stance – to know where someone is coming from. At first, most of us treat others as individuals rather than complex multitudes though, of course, when you spend enough time with someone, you become more aware of the complexity of who they actually are.

Where does the sense of self come from? To answer that, developmental psychology is an approach that seeks to explain the complexity of human psychology as an emerging and changing process. Some aspects of our identity are biologically well-established, encoded in our genes, such as eye colour or sex, and relatively uninfluenced by changing environments, though not immutable. Others are more flexible and unfolding, reflecting the interaction between biology and environments. Even then, the truth is that all biology operates in environments, so that the trite nature–nurture division, while popular with headline sub-editors, is illogical from a biological perspective. Yet, we continually talk about someone's true nature because it is such a powerful way of categorising others. Whether it is giftedness, kindness, intelligence, or humour, we are prone to consider these as intrinsic properties or facets of someone's self.

For most of us, the self is more than a list of attributes. A useful division one can make when considering the nature of the self is the distinction between 'I' and 'me' proposed by William James (1890). The phenomenological stream of consciousness that we experience is what James called the 'I experience'. But there is another type of self that is more familiar when we ask a stranger to 'Tell me about yourself.' What we expect to hear in response is personal information related to their experiences, attitudes, or education. William James referred to that common sense of self as the identity that we call 'me'.

As a developmental psychologist, I think the experience of 'I', the stream of consciousness, is probably present very early. I have no reason to doubt that babies experience consciousness. What they are experiencing and interpreting, of course, might be very different from that of an adult. However, the other Jamesian sense of self, the 'me experience', must be constructed as they enter different phases of their lives and come under the influence of events and others that shape who they become. I think that this 'me experience' must be changing radically over a lifetime. I would further conjecture that this sense of self does not just change in the individual. I have argued that sense of self has been changing in us as a species over the course of our evolution.

FIGURE 4.3 Charles Darwin.

This now brings me to Charles Darwin (Figure 4.3). For me, it is very gratifying as a psychologist that one of the world's greatest scientists recognised the importance of my field of study. On the last pages of *The Origin of Species* he writes:

> In the future I see open fields for far more important researches.
> Psychology will be securely based on the foundation already well laid out
> by Mr Herbert Spencer, that of a necessary acquirement of each mental
> power and capacity by gradation. Much light will be thrown on the origin
> of man and his history.

(Darwin, 1859, p. 402)

Darwin is saying that, if you really want to understand humans, it is not enough to just think about their development and evolution as adaptations to a physical environment, you must understand how their behaviours and thought processes have also been selected. Those environments are not just physical. They also include social environments.

Our social environment has undergone radical change, and I believe that it has undergone a significant change quite recently, in fact as recently as the last 20,000 years. I think that our species has undergone a significant degree of domestication. When one hears the word 'domestication' one usually thinks about the conveniences of living in modern societies, but the word 'domestication' has a much older origin, and, in fact, Darwin talked about domestication extensively in *The Origin of Species*. The book is, of course, based on his observations collected famously on his travels to the Galápagos Islands, but he spent a lot of the time talking to plant and animal breeders about the process of domestication, whereby they were selecting the attributes that were desirable and propagating them in the offspring. This gave Darwin the clue to the mechanism of natural selection to explain how diversity in life on Earth could emerge as adaptations to changing environments.

I think that humans have been changing because the environment, which has undergone the most significant change, has become a critically important social environment, and that happened fairly recently. This thesis I developed in my last book, *The Domesticated Brain*, and it starts with a really odd fact that you may not be aware of (Hood, 2014). If you look at the fossil record, skull sizes have been increasing over the course of our evolution. This reflects increasing brain size related to cognitive flexibility or intelligence but also an animal's capacity for social interaction. However, about 20,000 years ago, at the end of the last Ice Age, the skull size of modern humans shrank significantly, by about 10–15% (Stringer, 2011). That is about the size of a tennis ball, which is not a trivial amount of brain to lose. Now why would that happen?

There are a number of possible reasons and factors which are not mutually exclusive. First, at the end of the last Ice Age, of course, the planet warmed up, and this may have produced environments that support smaller-brained humans. However, as Chris Stringer, the palaeontologist, pointed out, there have been other climate changes throughout the course of our evolution that are not correlated with any reduction in brain size. There was also a massive increase in the human population around this time. We know this by analysis of the variation you get in the mitochondrial DNA that tells us that suddenly the human population exploded at about this time (Zheng *et al.*, 2012). We also changed lifestyles; we moved from being hunter-gatherers to settling down into much larger communities. All of these changes could have played some role in a significant reduction in our brain size. Yet, there is one more hypothesis, which seems too far-fetched, but worth considering. Brains shrink when species become domesticated through a process of socialisation.

Humans had been living in groups much earlier than 20,000 years ago. There is plenty of archaeological evidence that they even had rituals and religions for at least 50,000 years. However, at the end of the last Ice Age, this social environment needed a different type of psychology that needed a different type of mind. It needed one that was conducive to living in large collective groups, collaborating in a way that was much more integrated and complex than ever before. We began to select amongst ourselves for those individuals who had the psychological attributes that were the most conducive to living in this new environment. We preferred to breed with those who were psychologically better adapted to the new settled communal living which required social skills to negotiate and cohabit by sharing resources. Around 12,000 years ago, agriculture began to appear where humans domesticated wild animals and cultivated crops. Over that interim period of around 8,000 years, we had to develop a level of pro-sociality well before we could even begin to undertake the challenges of farming that requires collective efforts, mutual goals, for-ward planning, and trust. When we made the transition from nomadic hunter-gatherers to settled farmers who had to work together in order to plan the harvest, we had to become a radically different type of person. In effect, we began to self-domesticate ourselves as well.

Is there any evidence for this speculative notion? Remarkably, it turns out that, if you select for pro-sociality, you can change not only the psychology, but also the physiology of the offspring. This discovery primarily comes from the work of Dmitri Belyaev, a geneticist who was working in the Soviet Union in the 1950s. The Communist Party outlawed the study of genetics for its political overtones, and so Belyaev had to conduct his research out of the watchful eye of Stalin's secret police. So he took himself off to Siberia, where there were fur farms established to supply the demand for animal skins, to continue his studies. One species that was trapped and housed in these farms was the Siberian silver fox, one of the few remaining wild species, among animals whose fur is used by humans, that had never been tamed.

Belyaev introduced a programme of domestication of the Siberian silver fox whereby he selectively bred individual animals that were the most passive when they were approached. Remarkably, within only 20, or so, generations of this selective breeding, Belyaev produced a completely domesticated silver fox that was significantly different from its undomesticated wild cousin. Not only did these domesticated foxes behave differently, they looked differently. They had bushy tails and floppy ears (a feature that Darwin also noted in other domesticated animals), and their way of thinking was different.

Brian Hare, a comparative psychologist from Duke University, has looked at a number of wild and domesticated species, and collected evidence for a domestication syndrome that includes (1) physiological changes related to aggression, such as reduced reactivity of the hypothalamo-pituitary–adrenal (HPA) axis; (2) behavioural changes including reduced aggression and increased tolerance, and also increased pro-social behaviours, particularly play, nonconceptive sexual behaviour, and grooming; (3) cognitive changes including differences in problem-solving abilities, with more cunning in the wild animals but more dependency in the domesticated animals; and finally (4) reduced brain size that may be attributable to feminisation (Hare, 2007). Domestication also comes with a process known as juvenilisation whereby offspring take much longer to reach maturity and are much more dependent on caregivers. In fact, when some of Belyaev's foxes broke out of the farms and escaped into the wild, they came back very soon after because they could not survive by themselves (Trut *et al.*, 2009).

It is quite clear that selecting against aggression and for pro-sociality produces profound and lasting physical changes. When wild animals are domesticated, their bodies and brains change along with their behaviour. It is no surprise, then, that the brains of all the roughly 30 species of animals that have been domesticated by humans have decreased in volume by about 10–15% in comparison with the brains of their wild progenitors – the same reduction observed over the last 1,000 generations of humans.

I think domestication has been happening to us as a species as we selected against aggression and chose pro-sociality. Of course, this does not mean that aggression was wiped out, as aggression was necessary for competition, but whereas that competition may have been more at the individual level for small groups, larger populations and emerging societies needed to constrain that violence within the domestic setting. Ostracism and punishment by the group would have become the necessary mechanisms to control how we behaved in order to reach the age of puberty and produce offspring who would be accepted by the group. I would further speculate that maybe our childhood suddenly increased at the end of the last Ice Age and now we had a brain which took much longer to reach maturity by encoding more and more information, which made it much more adaptive to living in social groups.

To encode more information from those around us, we need a brain that is complex enough for social learning. The brain of any animal is as complex as it needs to be to solve the world problems that the creature has evolved to face. In other words, the more versatile an animal's behaviour, the more sophisticated its brain. This versatility comes from the capacity to encode information – storing memories as patterns of electrical connectivity in the specialised brain cells called 'neurons' that alter in response to experiences. In the human adult, the brain is comprised of an estimated 170 billion cells, of which 86 billion are neurons (Azevedo *et al.*, 2009). The neuron is the basic building block of the brain's communication processes that support thoughts and actions.

These neurons sit in the 3–4-mm-thick outer layer called the cortex processing and encoding information. This information initially comes in through our senses and is transduced into nerve impulses that traverse these vast networks of neurons connected by transmitting structures made up of axons (the long thin fibre), dendrites (shorter branching structures), and synapses (terminal channels between communicating cells). When the sum of incoming nerve impulses reaches a critical threshold, the receiving neuron then discharges its own impulse down its axon to set off another chain reaction of communication. As they fire, the weighting between the connected neurons changes, and in this way we can take information from the outside world and encode it in patterns of brain firing. Each neuron effectively acts like a miniature microprocessor.

The patterns of nerve impulses that spread across the vast network of trillions of neural connections are the language of the brain as information is received, processed, transmitted, and stored in these networks. The presentations of experiences become re-presented, or representations – neural patterns that reflect experiences and the internal computing processes our brains perform when interpreting information. Representations then feed back into the system to streamline and give context to further processing – in other words, knowledge.

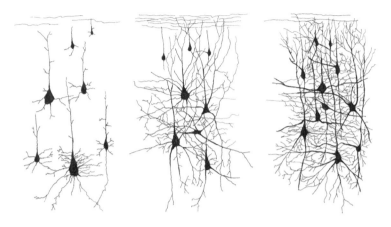

FIGURE 4.4 Increase in connectivity between cortical neurons with age from newborn to the first year of life.

What might surprise you is that babies are born with almost a full complement of neurons. They have just as many as adults if not more, about 86 billion neurons. But what they lack is all the connections that join them up. As illustrated in Figure 4.4, cortical neurons undergo considerable changes in their connectivity in early development.

In general, there are two processes driving this change in connectivity. First of all the cells are programmed by intrinsic mechanisms to form an exuberant over-production of connections. This produces increased myelination (insulation of the axons), arborisation (increase in dendritic density), and synaptogenesis (increase in synaptic channels). However, whether these connections are maintained or not depends on experience because, if neurons are not firing in synchrony together, the connections are lost. Two neurophysiological dictums capture this as 'Cells that fire together wire together' and 'If you do not use them, you lose them.' In this way, through these generative and degenerative processes, the brain becomes shaped by experience in a process known as 'plasticity'.

Plasticity reflects the importance of environmental input. It occurs throughout our lifetimes as it is the process that supports learning. Some aspects of plasticity occur within certain time frames known as 'sensitive periods'. The Nobel Prize-winning work by Hubel and Wiesel demonstrated that if you raise animals in abnormal visual environments you can produce permanent impairments that you cannot correct later on in life. So there is something critical about having a normal sensory experience early on for this particular function.

However, developmental psychologists are increasingly beginning to understand that what is true of the sensory systems of vision is also true of the social systems.

The human brain has a number of circuits that seem to be specialised for socially relevant jobs. For example, the lateral fusiform gyrus is related to face processing (Kanwisher *et al.*, 1997), whereas other areas including the superior temporal sulcus register biological movement (Grossman *et al.*, 2000). The anterior frontal area is active during mentalising when you try to imagine what somebody else might be thinking (Saxe, 2006). Then you have regions in the more temporal lobes including the amygdala and insula, which form part of the emotional circuitry registering those positive and negative things that we feel during social interactions (Lamm & Singer, 2010). There are regions in the premotor areas of the cortex that are active when we are watching people do things, which may support our human skill at mimicking and synchronising (Keysers, 2011).

Activity in all of these regions may present early in development, as rudiments for each of these social skills can be found before the first birthday even though we may not always have the ability to measure the corresponding brain activity in human babies. However, these skills become refined and honed with social experience over our childhoods. We have evolved an exceptional capacity to think and interact with each other, which enables us to live in complex social environments in a way that is unparalleled in the animal kingdom.

Humans have proportionally the longest childhood of any species on the planet. We expend a lot of our time investing our efforts in raising our children. This strategy is also found in other social animals, and it may be no coincidence that those animals also tend to be the more intelligent. However, I do not think we originally developed this extended childhood to instruct our children or educate them in the ways of the world. Education is probably a fairly recent invention in our history. Remember, dependency is a by-product of domestication. Rather I think long childhoods enabled us to become enculturated – to learn how to behave so that we can benefit from the knowledge and support of others. To do so, we have to become integrated with the group through socialisation. If you think about it, that is what we tell children all the time, 'Behave yourself!' Children need to learn the rules of how to behave so that they can benefit from society and go on to have their own children and learn from the collective group.

This ability to become enculturated requires early social engagement and social interactions, and may share some of the sensitive periods of plasticity that have been reported for sensory processing. For example, Harry Harlow raised baby rhesus monkeys in total isolation to see what happens if you give them everything but the social interaction. They were provided with water, warmth, and food, but they had no social inter-action. By varying the amounts of time that animals were kept in isol-ation, he discovered that if these monkeys spent longer than six months without any social interactions then they were permanently impaired in social behaviour. When they were reintegrated back into the troop they were ostracised. However, if the monkey was reintroduced within six months, then it could be fully rehabilitated and reintegrated back with the rest of the troop. Thus Harlow had demonstrated a sensitive period for social interaction in the rhesus monkey (Harlow, 1958).

For obvious ethical reasons, isolation experiments with human chil-dren are forbidden, but there are examples where the effects of early abnormal impoverished social environments can be observed. When Westerners arrived in Romania in 1990, after the collapse of Ceaușescu's regime, they discovered that he had been forcing women to have children they could not possibly look after, and so they were dumping them in State orphanages. Very often there were 40 children being looked after by just one caregiver. They did not have any social interactions whatsoever.

Many of these children were evacuated and fostered out to homes in the USA, Canada, and the UK. In the UK, Sir Michael Rutter, the famous psychiatrist, undertook a longitudinal study of a group of them to determine whether these terrible early experiences had left permanent damage on their development.

He took 111 of these children and then compared them with a matched sample of British adoptee children who had not had an early abnormal environment to act as a control group for comparison. Both groups were then assessed on a variety of measures followed up over the years in a longitudinal design. What Rutter discovered was that, despite an initial really significant impairment compared with the British control group on intelligence scores and other cognitive measures, the Romanian orphans quickly made up a lot of ground. There was much recovery despite the terrible beginning in the orphanages. However, that improvement was dependent on how long they had been isolated. Moreover, there was evidence that they remained impaired in their social development.

One could argue that the longer they had been in the orphanage, the more physically malnourished they were. However, when an analysis checking for initial body weight as a measure of malnutrition was conducted, it was not the physical state that was the best predictor of the child's outcome, but rather the amount of time they had been isolated (Rutter *et al.*, 2004). What was really surprising was that the critical period for isolation was also six months. Just like Harlow's rhesus monkeys, there seems to be something really important about forming stable relationships during the first six months that has profound consequences much later down the road. These Romanian orphans are now adults living in our culture, but they still have a legacy of problems. Most of them find it difficult to form social relationships, since they exhibit behavioural problems associated with anti-social behaviour.

What could possibly explain the long-term effects of early social deprivation? Well, I think the evidence strongly points to the fact that their reactivity to stress has been disrupted. We have a system called the fight–flight response, which is controlled by this hypothalamo-pituitary–adrenal (HPA) axis. The HPA axis is activated when you have to respond to a potential threat. It is the same system that is attenuated during the domestication process in foxes. What might be going on in this early developmental phase is that socially isolated or impoverished children are not developing regulatory systems; they are developing abnormal responses because they cannot predict what is going to happen next. Their HPA axis does not develop normal reactivity.

One of the characteristics of abnormal socially abusive early environments is that they are not necessarily aggressive. In fact, children might just be neglected and ignored, which is still a situation where they fail to learn how to regulate. The hypothesis that a lack of predictability affects our physiological response to stress is supported by animal studies. If you induce stress during early postnatal developments, the reactivity of the

HPA axis to stress is altered (Liu *et al.*, 1997). This can even occur in the prenatal period for monkeys (Clarke *et al.*, 1994).

Evidence of a similar relationship between the effects of early stress and later reactivity of the HPA axis is beginning to accumulate in humans. A recent study of Dutch pensioners who were evacuated during the war showed that they too have disrupted HPA sensitivities (Pesonen *et al.*, 2010).

Now let us turn to social development in the typical child. We are born to be sociable or, as Mike Tomasello has quipped, 'Fish are born expecting water, humans are born expecting culture.' When you look at very young babies, there is much experimental evidence that they are pre-adapted in many ways. Right from the very beginning, newborns prefer to look at faces (Johnson & Morton, 1991). They prefer the sound of the human voices and especially their own mother's (DeCapser & Spence, 1986). At two months, they will smile back when they are smiled at (White, 1985) and will join in shared attention (Hood *et al.*, 1998). These social behaviours are all easily triggered before the first six months of life.

People are the most interesting objects to babies. They pay attention to adults right from the start, and the adults they like the best are the ones who appear to be paying most attention to them (Roedell and Slaby, 1977). The way they figure this out is by looking for synchronised contingent behaviour. If an infant makes a response and the adult responds contingently, then the child prefers that adult as opposed to another adult who does not pay them any attention. In fact, if you ignore the infant, as in the classic 'still-face' paradigm, where the mother stares blankly at their infant, this causes much distress to the child (Ellsworth *et al.*, 1993).

During these early social interactions, I think what the infants are really doing is trying to figure out who is more interested in them. For example, consider imitation in very young children where they copy an adult. It has often been thought that imitation must be a mechanism of learning by observation. This is one reason why researchers study non-human animals looking for evidence of imitation. When young primates can observe others cracking nuts with a stone hammer or prodding termite hills with rudimentary twig tools, this is interpreted as tool use and evidence of learning by observation.

Humans readily imitate others, but I do not think that learning by observation is why we imitate as infants. Rather, when an infant imitates an adult, most of us typically find that mimicry endearing. There is something very cute and funny about infants acting like adults, and so we show strong positive emotions towards them. I think this is their strategy. They are tugging at our heartstrings by making us feel emotionally committed and bonded to them. That is the real role of imitation. They say that imitation is the most serious form of flattery. I think that these little guys are deliberately imitating in order to trigger in us a bonding social response.

Now, not only do they start to pay attention to which adults are responding to their clowning around, but also they want to know who are the good guys and who are the bad guys. This is a research programme which originated with my colleagues Paul Bloom and Karen Wynn at Yale University. One of their former graduate students, Kiley Hamlin, has been looking at the beginnings of moral reasoning using plays with puppets behaving either pro-socially or anti-socially. In one study, infants watch a hand puppet trying to open a box without much success, until a second puppet assists the first one to lift the lid. The infant then watches another version where again the puppet is trying to retrieve the object, but this time a different puppet jumps on top of the lid of the box, thwarting the original puppet's efforts. Having just witnessed these two characters, the infant is then offered the opportunity to interact and choose which one of the dolls they like the most. Before the age of one year, infants preferentially choose the one that they have seen being helpful (Hamlin *et al.*, 2011). What is true of hand puppets is also true of humans. When they see adults behaving either pro-socially or anti-socially they also have preferences towards the one whom they think is going to be the most helpful (Dunfield & Khulmeier, 2010).

When babies grow up to become toddlers who are motorically capable of helping themselves, they start to initiate activities to engage the pro-social behaviour to help others. If an adult 'accidentally' drops a pencil or is unable to open a door, toddlers will come to their assistance (Warneken & Tomasello, 2006). Again, I think children are being strategic. They are engaging in behaviours deliberately to engender pro-sociality and our responses back to them. They are seeking to identify those adults who are most likely to help them. This could be considered a strategy of promiscuous altruism – toddlers are just kind to everyone, but very soon afterwards children show a sort of darker side to their nature because they start to show what we call in psychology 'in-group' preferences and 'out-group' biases or prejudices. When they identify with others they will share their own resources with them, but if they see someone as an out-group member then they're decidedly very anti-social towards them.

In order to recognise both in-group and out-group members, you have got to have a sense of which group you belong to. In other words, you have to begin to understand your own self and your own self-identity, and that has different developmental milestones over the years.

One of the classic markers on the road for self-identity is the self-recognition mirror test or the rouge test first developed by Gordon Gallup following the observations of Darwin. On visiting London Zoo, Darwin noted that many animals when presented with their own reflection would often attack the mirror as if the image were another animal. On the basis of this observation, Gallup developed a technique where he put red makeup on the nose of the animals when they were asleep and then presented them with a mirror to see how they responded (Gallup, 1970). Typically, many animals respond as if the reflection were another animal, except, notably, for those that live in social groups, such as elephants and dolphins. In humans, infants generally will not pass the rouge test until about 18 months of age. Until that age they just think the reflection is another baby and laugh at it. Around about 18 months of age they will regard themselves in the mirror, stare at the mark, and then touch their own nose to remove it, indicating that they recognise themselves in the mirror.

Memory researcher Mark Howe thinks that babies who fail the rouge mirror test lack a sense of self, which he argues has important implications for forming autobiographical memories (Howe & Courage, 1993). This might explain the phenomenon of infantile amnesia, which is the inability to recollect personal memories before the second birthday (Tustin & Hayne, 2010).

In order for memories to possess meaning, they have to be embedded within an enduring sense of self. It used to be thought that long-term memories cannot form until brain structures, including the hippocampus, undergo maturation. However, research also supports the perspective that, without a coherent sense of self, experiences cannot be integrated into a meaningful narrative that forms the basis for the majority of autobiographical memories. This may explain why Asian cultures have earlier autobiographical memories compared to those in Western cultures (Wang, 2006). When parents talk to their children and review the day's events, children can encode the events within a self framework provided by the adult.

Another milestone in the construction of self-identity is gender. As parents (and often grandparents), one of the first things we get fixated on is 'Is it a boy or is it a girl?' Even before infants are born, some parents will paint the nursery either blue or pink or start buying gender-appropriate clothes. This notion of gender identity is something that children become really quite obsessed with, because they start to look out for all the examples for the gender that they believe they are supposed to belong to.

Self-labelling as a boy or girl is one of the first markers of identity, and, by the time they are two years old, most children have a preference for their own gender (Maccoby & Jacklin, 1987). In fact, sensitivity to gender predates racial prejudice, which appears much later. When asked to select potential friends from photographs, three- and four-year-olds show a reliable preference for their own gender, but not their own race (Abel & Sahinkaya, 1962).

Once they know they are a boy or a girl, they become gender detectives, seeking out information about what makes boys different from girls (Miller *et al.*, 2013). This is when they begin to conform to the cultural stereotypes present in society. Not only are they gender detectives, but they also police the differences as enforcers, criticising those who display attitudes or behaviours associated with the opposite gender. They start to say things like 'Boys cannot wear pink' or cannot have long hair or cannot wear dresses, and they try to enforce this to strengthen the in-group and out-group identity. By three to five years of age, children are already saying negative things about other children whom they do not identify with. They are making a distinction between in-groups and out-groups. If you are in my gang, then we are both in-group members.

Egocentrism is another important aspect of a child's behaviour that must be controlled if you are to develop a mature sense of self. This work comes from Jean Piaget, the famous Swiss child psychologist who noted that children have a very egocentric view of the world. They think that everyone thinks just the same things as they do and sees the world in just the same way they do. Without the capacity to take into account another person's perspective, social interactions are going to be fairly limited. You have to understand that other people see the world differently from you. Egocentrism is also related to the next major milestone in developing a sense of self for social interactions called a 'theory of mind'. This is the capacity to understand that not only do people have minds different from your own, but they may think things are different or they might hold a different view. They may even hold a false belief that they believe to be true, but you know to be wrong. For example, in a classic measure known as the 'Smarties Task', children are shown a box of Smarties® and asked 'What do you think is inside?' Typically, children aged three and four years reply 'Smarties'. The box is then opened to reveal the contents, which are actually pencils. Now when asked to say what they thought was in the box, typical four-year-olds recognise that they had a false belief and respond 'Smarties', whereas the younger three-year-olds respond 'Pencils'. Moreover, if you then ask the child what their friend will answer, who does not know the true contents of the box, again four-year-olds will understand that their friend too will hold a false belief, whereas the three-year-olds will answer 'Pencils'. So they fail to understand the possibility of false belief, namely that people can be mistaken (Gopnik & Astington, 1988). It is as if they cannot ignore what they know to be true, which has been called the 'curse of knowledge' (Birch & Bloom, 2007).

Other measures of false belief include the Sally-Anne task, where children have to work out whether characters in a story understand when they have been deceived (Wimmer & Perner, 1983). Again these measures of theory of mind reveal that, between three and four years of age, children make a conceptual transition in understanding how others think. An inability to easily understand another person's mental state would be a real impairment in social interactions, and it is noteworthy that individuals with autism spectrum disorder often fail these mentalising tasks (Baron-Cohen et al., 1985). If you do not know what another person is thinking, you cannot easily predict their reasoning. Moreover, with a theory-of-mind skill, you can be Machiavellian and deliberately deceive or manipulate others to your own advantage.

When children become aware that other people have thoughts and opinions, one of the first concerns that children start to become obsessed with is 'What do they think about me?' This is where our self-esteem begins to become an important component of our self-construction, and indeed it remains so for the rest of our lives. For example, in our lab, we've been looking at the use of ownership as a measure of self-esteem. This research builds upon the musings of William James about the nature of self when he wrote:

> *In its widest possible sense*, however, *a man's Self is the sum total of all that he* CAN *call his*, not only his body and his psychic powers, but his clothes and his house, his wife and children, his ancestors and friends, his reputation and works, his lands and horses, and yacht and bank-account.
> (James, 1890, p. 292)

The quote is a bit archaic and misogynist, but what James is saying is that we use possessions to signal our status and we covet objects of value because this gives us a sense of self-worth and self-esteem. This is also true of children. Children will desire objects that they think other children want as well. About 75% of young children's conflicts with peers revolve around the possession of objects (Shantz, 1987), and toys are more desirable amongst pre-schoolers when they have been touched or named by another child, which is consistent with the hypothesis that ownership operates as a status mechanism (Hay & Ross, 1982).

The flipside of ownership is sharing. Generally, young children have a tendency to maximise self-gain, and it is not until they are aged five years and older that they begin to share resources equally (Blake & Rand, 2010). Up until that point they are incredibly selfish. In various sharing games, pre-schoolers are selfish unless the other child has helped them to receive rewards (Hamann *et al.*, 2012). From about six years of age, children will share, but they are still strategic as they are biased towards others whom they identify with (Fehr *et al.*, 2008). Sharing may be a pro-social and desirable trait, but it is one that has to be strategic. You need to share your resources with those in the in-groups who are likely to help in the future and avoid those in the out-groups who would not feel obligated to reciprocate.

The final important milestone I would like to consider in the path to developing the self is self-control. In most of the tasks I have mentioned up to now, success depends on regulating and controlling thoughts and impulses so that one is less egocentric. If you cannot regulate and control your behaviours, then that can make social interactions very difficult. You need to be able to withhold or delay your own gratification, which requires inhibitory mechanisms to suppress intrusive thoughts and actions.

A child's developing capacity to self-control has been made famous in Walter Mischel's 'marshmallow test' (Mischel *et al.*, 1972). In this task children are presented with a tasty marshmallow on a plate in front of them and told 'Do not eat the marshmallow because, if you wait, you can have two when I [the researcher] come back into the room.' This turns out to be quite difficult for young children, who either cannot wait or fidget in an attempt to distract themselves. Mischel has demonstrated that a child's capacity to delay gratification at age four predicted how well they performed academically at school 11 years later, how well-adjusted they become as adults, and even whether they turn into drug addicts later in life (Mischel *et al.*, 1989, 2011). It also predicted their social behaviour, how many friends they had, and also how stable they were in their emotional relationships.

One reason why self-control is so important for relationships is that, when you are engaged in social interaction, it requires a bit of give and take. We all know people who are so egocentric or impulsive that they cannot regulate their own behaviour and will either dominate a conversation or interrupt others. You have to learn to control your own urges and impulses in order to have satisfying and productive social interactions.

The brain is a multitude of developing systems, coordinated and integrated to produce this sense of self. All of these components of the developing self I have discussed so far are supported by integrated neural systems that mature at different rates. The prefrontal lobes, which are partly to do with inhibiting behaviours, enabling us to plan and to mentalise what others are thinking, undergo prolonged development during the course of childhood (Blakemore, 2008).

We have little insight into the neural parallel processing which underlies our stream of consciousness. We feel that we are the ones in control

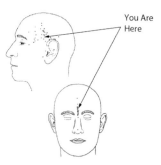

FIGURE 4.5 Plot of the locations that people report as the seat of their conscious awareness. (After Bertossa *et al.* (2008).)

as an integrated coherent individual self but, as I have argued in my book *The Self Illusion* (Hood, 2012), that experience is not what it seems. I would argue that the neuroscience indicates that there is no individual at the core of our identity, but rather we are a multitude of experiences that arise both from the representations stored in our brains and from the influences and external factors that compete to drive our system. And yet we feel we are inside our head like a little puppet master.

As my colleague Paul Bloom (2005) has noted, 'We do not feel that we *are* our bodies. Rather, we feel that we *occupy* them, we *possess* them, we *own* them.' We feel we are operating a sophisticated machine, situated somewhere inside our heads located behind the eyes. This is the common phenomenology that most of us experience in our daily consciousness. In one study people were asked to introspect, to look in upon their consciousness in the way that some forms of meditation work, to try to locate their present seat of consciousness (Bertossa *et al.*, 2008). They were asked to shut their eyes, relax, and then point to where they felt the origin of their consciousness was coming from. As the plot of points in Figure 4.5 indicates, people pointed to their head and felt they were situated somewhere behind their eyes. The notion of a smaller version of our self, located inside our head is known as the homunculus (Latin 'little man'), but explaining our experiences by evoking such a creature creates a logical problem. If there were a homunculus in control of our thoughts and actions, and who was the experiencer at the centre of our sense of self, then who is inside the head of the homunculus? It cannot be another homunculus because the same problem applies. The homunculus problem as it has become known is one of an infinite regress.

Rather, the brain is a massively parallel system integrating the output from relatively separable systems that is somehow brought together as the experience of self. This sense of self is most often, but not always, coherent, but, if you disrupt the communication between the various brain regions, then one can see how the self can fractionate.

One of the classic examples of this fractionation comes from the work of Michael Gazzaniga, who studied patients who had intractable epilepsy and required an operation to disconnect the two halves of the brain, by cutting through the bundle of fibres known as the corpus callosum, which generally shares information between the two hemispheres. After the operation, these 'split-brain' patients have effectively two brains operating in isolation from each other.

In one of his studies (Gazzaniga et al., 1962), he presented split-brain patients with pictures and words on a computer screen in the left and right visual fields. For example, if they saw the word 'ring' on the right side of the screen and the word 'key' was presented on the left side, both words would be processed separately in the opposite hemisphere. Owing to the way the anatomical connections cross over, all the information on the left side of the world is processed by the right hemisphere and all the information and actions on the right side of the world are processed by the left hemisphere. Language speech production is typically in the left hemisphere, so if you then ask split-brain patients to say what they see on the screen, they say the word in the right field, which is 'ring'. However, if you ask patients to pick up the object with their left hand, which is controlled by the right hemisphere, then they pick up the key.

In another example of the strange case of spilt-brains, Gazzaniga (2005) gave one of his patients a blocks puzzle to solve using only his right hand (controlled by the language-dominated left hemisphere). However, this was a spatial puzzle where the blocks had to be put in the correct position (something that requires the activity of the right hemisphere). With the right hand, the patient was hopeless, turning the blocks over and over until, as if frustrated, the left hand, which the patient had been sitting on, jumped in and tried to take the blocks away from the patient's right hand. It was as if the hand had a different personality. In another observation, a naked man was flashed into the right hemisphere, causing a female split-brain patient to laugh, but not be able to say what it was she was finding amusing. Her left hemisphere was unaware of the naked man and so could not explain what was amusing.

When patients were presented with these inconsistencies between what they said and what they did, Gazzaniga noted that the patients would confabulate a story to make sense or to reconcile the differences in their behaviour. Gazzaniga further proposed that there is a system he called the 'interpreter', which enables us to try and piece together all these inconsistencies to make sense of everything. However, you do not have to have your brain split in two to demonstrate how easily we generate explanations to make sense of inconsistencies in our thoughts and actions.

In a study of an attentional phenomenon called 'choice blindness' (Figure 4.6), normal participants were shown pairs of faces and asked to choose the one they preferred (Johansson *et al.*, 2005). After making a choice, the participant was handed the photograph and asked to justify their decision. On some of the trials the faces were surreptitiously switched by sleight of hand, so that the participant ended up with the face they had just rejected. Remarkably, if they did not notice the switch, participants went on to give full explanations of why this face was better than the other even though they had actually initially preferred the other. Choice blindness also works for various tests of food preferences where we initially reject a flavour and then give valid reasons for why it is better after we have been fooled by the experimenter with a switch.

What choice blindness suggests is that we often post-rationalise a lot of our experiences, which is why the psychologist Steven Pinker has quipped that consciousness is not so much the captain in control but more of a spin doctor of experience. In many instances we do not and cannot attend to everything and just assume continuity. We make our decisions often at the unconscious level and then apply conscious appraisal to rationalise the choice. Even when there are inconsistencies in the process, we smooth them over to maintain a consistent story. This process may also explain probably one of the most famous forms of distortion and bias in psychology, known as cognitive dissonance. Although cognitive

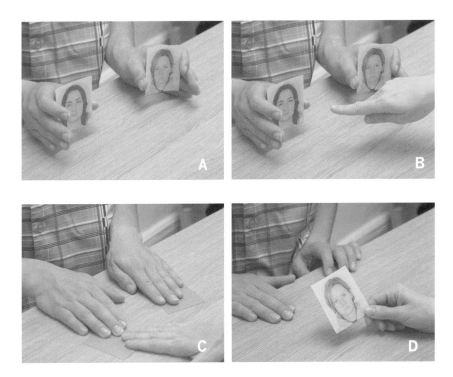

FIGURE 4.6 'Choice blindness'. Participants were shown two faces (a) and asked to choose the one they preferred (b). This was then passed to them (c), and they were asked to justify their choice. Every so often, the faces were switched (d). (After Johansson *et al.* (2005).)

dissonance has been around since the time of Aesop when he wrote the tale of the fox and sour grapes, Leon Festinger (1957) studied cognitive dissonance extensively in the 1950s. If you remember Aesop's fable, the fox wanted the grapes that were out of reach, but, on failing to reach them, the fox decided that they were probably sour anyway. This is where we get the phrase that someone is acting out of 'sour grapes'. It is not an unconscious process, but one where we feel compelled to reconcile inconsistencies in our self-image and outcomes.

Cognitive dissonance happens when we reframe events to maintain a positive bias or image of who we are rather than deal with the fact that we are maybe not as good as we think we are. If we put in a lot of effort to no avail, then, rather than accepting that we have failed, we say that we did not want the goal in the first place. We interpret our failure to attain a goal as actually turning out to be a good thing because, with hindsight, we reinterpret the goal as not really desirable. Otherwise, we are faced with the prospect that we have wasted a lot of work and effort to no avail. This discrepancy creates the cognitive dissonance. It is a dissonance because, on the one hand, we believe that we are generally pretty successful at attaining our goals. On the other hand, we were unsuccessful at achieving this particular goal. This is the dissonance aspect of our reasoning – the unpleasant mental discomfort we experience. To avoid the conflict this dissonance creates, we reinterpret our failure as a success. We have all done this to some extent where we put effort into trying to get a job or trying to secure a relationship. When we fail, we tend to say that we did not really want the job or the relationship.

The need to maintain a coherent self can also be found in many other famous examples from the realms of social psychology from the same era. For example, people think that they would be quite reliable and not be influenced by others, but we do know that, if you have enough people in a room saying that black is white, then there is a lot of pressure not to disagree with them. Solomon Asch famously showed that most of us tend to comply and conform with the group consensus (Asch, 1956). In a typical Asch conformity study, he held up cards with lines on them and went round the room asking eight participants which line matched the test line. In fact, there was only one real subject as the other seven participants were actually confederates of the experimenter. At first, everything seemed above board. Everyone agreed on the length of the test line on the first two trials. However, on the third trial, all of the confederates gave the wrong answer, stating that two lines that were clearly mismatched were equal in length. On average, three out of every four participants went along with fellow participants and also gave the wrong answer. Each did not suddenly become blind, but rather conformed in accordance with the group so as not to be the outsider. Each participant was fully aware of the correct answer, but did not want to appear different. They did not want to be ostracised, so they conformed to the group consensus.

This need to fit in and belong is part of the explanation in the last set of classic social psychology experiments I want to discuss – the infamous studies of Stanley Milgram. Like many other post-war social psychology researchers, Milgram, a student of Asch, wanted to know how ordinary citizens could do such cruel things to their countrymen as had been witnessed in Germany during WWII. Ordinary members of the public were invited to take part in a psychology experiment to investigate the effects of punishment on learning at Yale University. When each of the volunteers arrived at the laboratory, they were met by the experimenter, wearing a white lab coat, and another middle-aged man, who was introduced as another participant, but who was actually a trained actor. After a supposedly random decision, the experimenter explained that the volunteer would play the role of teacher and the actor would play the role of learner. The learner was led off to another room and it was explained that the teacher would read words to the learner over an intercom. The learner would then repeat the words back to the teacher. If the learner made a mistake, the teacher would press a button that delivered an electric shock to the learner in the other room. There were 30 levels of shock, rising in 15-volt increments from a safe level to one that was marked by a skull and an ominous triple 'XXX'.

Initially everything was going well, until the learner started to make mistakes and the true participant was asked to administer increasingly higher and more dangerous electric shocks. The shocks and the screams of the actor were bogus of course, but the participants believed they were inflicting pain and asked to stop the experiment. Nevertheless, around two out of three participants continued administering shocks even after there were screams followed by silence coming from the next room (Milgram, 1963). There is continuing controversy about these experiments and times have changed since the post-war period, but the point is still valid that most of us will behave in ways which seem to defy our representation of who we think we are, as an integrated coherent self.

The human species underwent an incredible change 20,000 years ago when the ice sheets began to recede. Our social environment changed in ways that forced us to become domesticated and more sociable. As a species, we have benefited from this intense social environment that we have created and the collective wisdom that enabled us to build the technologies and generate an exponential understanding of the world around us. However, the flipside of all this pro-social behaviour is that we have become increasingly co-dependent to the extent that we have a modern problem of loneliness.

Ostracism is one of the worst things you can do to another human being. Nelson Mandela noted this in his autobiography, *The Long Walk to Freedom*, when he talked about the corrosive effects of ostracism in fellow prisoners who would prefer to be tortured rather than left in total isolation. Mandela himself wrote that he had conversations and social interactions with insects because he was so lonely. It is so important for us to be included that we would rather suffer physical pain than be left alone and that when there is no one else around we will create them in this need to have social interaction.

Loneliness is not just a psychological problem. A recent meta-analysis combining the results of 148 studies and over 300,000 participants found that people with stronger social relationships had a 50% greater likelihood of survival than those with weaker social relationships (Holt-Lunstad *et al.*, 2010). Being alone was as much a health hazard as lack of exercise, moderate smoking, drinking, and obesity. In the UK today, it has been estimated that at least seven million people live alone – a terrible ticking time bomb if you like for the problems of social isolation.

In conclusion, in the same way that the Kanizsa triangle is not really there, but it is there because it emerges out of the context of everything around it, I think we can consider the development of the emerging self out of the context that has shaped us over our lives through the interactions that we have. All of these experiences have helped to form us, but if you take these away, maybe through the effects of isolation or through the effects of brain disorders where people lose their connections and representations of who they are, you can see that sense of self eventually disappearing until it goes.

Select Bibliography

Abel, H., & Sahinkaya, R. (1962). Emergence of sex and race friendship preferences. *Child Development*, 33(4), 939–943.

Asch, S. E. (1956). Studies of independence and conformity: I. A minority of one against a unanimous majority. *Psychological Monographs: General and Applied*, 70, 1–70.

Azevedo, F. A. C., Carvalho, L. R. B., Grinberg, L. T., Farfel, J. M., Ferretti, E. E. L., Leite, R. E. P., Jacob Filho, W., Lent, R., & Herculano-Houzel, S. (2009). Equal numbers of neuronal and nonneuronal cells make the human brain an isometrically scaled-up primate brain. *Journal of Comparative Neurology*, 513, 532–541.

Baron-Cohen, S., Leslie, A. M., & Frith, U. (1985). Does the autistic child have a 'theory of mind?' *Cognition*, 21, 37–46.

Bertossa, F., Besa, M., Ferrari, R., & Ferri, F. (2008). Point zero: A phenomenological inquiry into the seat of consciousness. *Perceptual and Motor Skills*, 107, 323–335.

Birch, S. A., & Bloom, P. (2007). The curse of knowledge in reasoning about false beliefs. *Psychological Science*, 18(5), 382–386.

Blake, P., & Rand, D. (2010). Currency value moderates equity preference among young children. *Evolution and Human Behavior*, 31, 210–218.

Blakemore, S.-J. (2008). The social brain in adolescence. *Nature Reviews Neuroscience*, 9, 267–277.

Bloom, P. (2005). Is God an Accident? *The Atlantic*. December Issue. www.theatlantic.com/magazine/archive/2005/12/is-god-an-accident/304425/.

Clarke, S., Wittwer, D. J., Abbott, D. H., & Schneider, M. L. (1994). Long-term effects of prenatal stress on HPA axis activity in juvenile rhesus monkeys. *Developmental Neurobiology*, 27, 257–269.

Darwin, C. (1859). *The Origin of Species*. London: John Murray.

DeCasper, A. J., & Spence, M. J. (1986). Prenatal maternal speech influences newborns' perception of speech sounds. *Infant Behavior & Development*, 9, 133–150.

Dunfield, K. A., & Khulmeier, V. A. (2010). Intention-mediated selective helping in infancy. *Psychological Science*, 21, 523–527.

Ellsworth, C., Muir, D., & Hains, S. (1993). Social-competence and person–object differentiation. An analysis of the still-face effect. *Developmental Psychology*, 29, 63–73.

Fehr, E., Bernhard, H., & Rockenbach, B. (2008). Egalitarianism in young children. *Nature*, 454, 1079–1084.

Festinger, L. (1957). *A Theory of Cognitive Dissonance*. Stanford, CA: Stanford University Press.

Gallup, G. (1970). Chimpanzees' self-recognition. *Science*, 167, 86–87.

Gazzaniga, M. S. (2005). Forty-five years of split-brain research and still going strong. *Nature Reviews Neuroscience*, 6, 653–659.

Gazzaniga, M. S., Bogen, J. E., & Sperry, R. W. (1962). Some functional effects of sectioning the cerebral commissures in man. *Proceedings of the National Academy of Sciences of the United States of America*, 48, 1765–1769.

Gopnik, A., & Astington, J. W. (1988). Children's understanding of representational change and its relation to the understanding of false belief and the appearance–reality distinction. *Child Development,* 59(1), 26–37.

Gregory, R. L. (1966). *Eye and Brain: The Psychology of Seeing.* London: Weidenfeld & Nicolson.

Grossman, E., Donnelly, M., Price, R., Pickens, D., Morgan, V., Neighbor, G., & Blake, R. (2000). Brain areas involved in perception of biological motion. *Journal of Cognitive Neuroscience,* 12, 711–720.

Hamann, F., Warneken, F., Greenberg, J. R., & Tomasello, M. (2012). Collaboration encourages equal sharing in children but not in chimpanzees. *Nature,* 476, 328–331.

Hamlin, J. K., Wynn, K., Bloom, P., & Mahajan, N. (2011). How infants and toddlers react to antisocial others. *Proceedings of the National Academy of Sciences of the United States of America,* 108, 19931–19936.

Hare, B. (2007). From nonhuman to human mind. What changed and why? *Current Directions in Psychological Science,* 16, 60–64.

Harlow, H. F. (1958). The nature of love. *American Psychologist,* 13, 573–685.

Hay, D. F., & Ross, H. S. (1982). The social nature of early conflict. *Child Development,* 53, 105–113.

Holt-Lunstad, J., Smith, T. B., & Layton, J. B. (2010). Social relationships and mortality risk: A meta-analytic review. *PLoS Med* 7, e1000316. doi:10.1371/journal.pmed.1000316.

Hood, B. (2012). *The Self Illusion: How the Social Brain Creates Identity.* Oxford: Oxford University Press.

Hood, B. (2014). *The Domesticated Brain.* London: Penguin.

Hood, B. M., Willen, J. D., & Driver, J. (1998). An eye direction detector triggers shifts of visual attention in human infants. *Psychological Science,* 9, 53–56.

Howe, M. L., & Courage, M. L. (1993). On resolving the enigma of infantile amnesia. *Psychological Bulletin,* 113, 305–326.

James, W. (1890). *Principles of Psychology.* New York, NY: Henry Holt and Company.

Johansson, P., Hall, L., Sikström, S., & Olsson, A. (2005). Failure to detect mismatches between intention and outcome in a simple decision task. *Science,* 310, 116–119.

Johnson, M. H., & Morton, J. (1991). *Biology and Cognitive Development: The Case of Face Recognition.* Oxford: Blackwell.

Kanizsa, G. (1979). *Organization in Vision.* New York, NY: Praeger.

Kanwisher, N., McDermott, J., & Chun, M. (1997). The fusiform face area: A module in human extrastriate cortex specialized for the perception of faces. *Journal of Neuroscience*, 17, 4302–4311.

Keysers, C. (2011). *The Empathic Brain: How the Discovery of Mirror Neurons Changes Our Understanding of Human Nature*. Amsterdam: Social Brain Press.

Laeng, B., & Endestad, T. (2012). Bright illusions reduce the eye's pupil. *Proceedings of the National Academy of Sciences*, 109, 2162–2167.

Lamm, C., & Singer, T. (2010). The role of anterior insular cortex in social emotions. *Brain Structure & Function*, 214, 579–591.

Liu, D., Diorio, J., Tannenbaum, B., Caldji, C., Francis, D., Freedman, A., Sharma, S., Pearson, D., Plotsky, P. M., & Meaney, M. J. (1997). Maternal care, hippocampal glucocorticoid receptors, and hypothalamic–pituitary–adrenal responses to stress. *Science*, 277, 1659–1662.

Loftus, E. F. (1975). Leading questions and eye-witness report. *Cognitive Psychology*, 7, 560–572.

Maccoby, E. E., & Jacklin, C. N. (1987). Gender segregation in childhood. *Advances in Child Development and Behavior*, 20, 239–287.

Meltzoff, A. N., & Moore, M. K. (1977). Imitation of facial and manual gestures. *Science*, 198, 75–78.

Milgram, S. (1963). Behavioral study of obedience. *Journal of Abnormal and Social Psychology*, 67, 371–378.

Miller, C. F., Martin, C. L., Fabes, R. A., & Hanish, L. D. (2013). Bringing the cognitive and social together. In M. R. Banaji and S. A. Gelman (eds.), *Navigating the Social World: What Infants, Children, and Other Species Can Teach Us*. New York, NY: Oxford University Press.

Mischel, W., Ebbesen, E. B., & Raskoff-Zeiss, A. (1972). Cognitive and attentional mechanisms in delay of gratification. *Journal of Personality and Social Psychology*, 21, 204–218.

Mischel, W., Shoda, Y., & Rodriguez, M. L. (1989). Delay of gratification in children. *Science*, 244, 933–938.

Mischel, W., Ayduk, O., Berman, M. G., Casey, B. J., Gotlib, I. H., Jonides, J., Kross, E., Teslovich, T., Wilson, N. L., Zayas, V., & Shoda, Y. (2011). 'Willpower' over the life span: Decomposing self-regulation. *Social Cognitive and Affective Neuroscience*, 6(2), 252–256.

Pesonen, A. K., Räikkönen, K., Feldt, K. K., Heinonen, K., Osmond, C., Phillips, D. I., Barker, D. J., Eriksson, J. G., & Kajantie, K. (2010). Childhood separation experience predicts HPA axis hormonal responses in late adulthood: A natural experiment of World War II. *Psychoneuroendocrinology*, 35, 758–767.

Rochat, P. (2009). *Others in Mind: Social Origins of Self-Consciousness.* Cambridge: Cambridge University Press.

Roedell, W. C., & Slaby, R. G. (1977). The role of distal and proximal interaction in infant social preference. *Developmental Psychology*, 13, 266–273.

Rutter, M., O'Connor, T. G., & The English and Romanian Adoptees (ERA) Study Team (2004). Are there biological programming effects for psychological development? Findings from a study of Romanian adoptees. *Developmental Psychology*, 40, 81–94.

Saxe, R. (2006). Uniquely human social cognition. *Current Opinion in Neurobiology*, 16, 235–239.

Shantz, C. U. (1987). Conflicts between children. *Child Development*, 58(2), 283–305.

Stringer, C. (2011). *The Origin of Our Species.* Penguin, London.

Trut, L., Oskina, I., & Kharlamova, A. (2009). Animal evolution during domestication: The domesticated fox as a model. *BioEssays*, 31, 349–360.

Tustin, K., & Hayne, H. (2010). Defining the boundary: Age-related changes in childhood amnesia. *Developmental Psychology*, 46, 1049–1061.

Von Der Heydt, R., Peterhans, E., & Baumgartner, G. (1984). Illusory contours and cortical neuron responses. *Science*, 224, 1260–1262.

Wang, Q. (2006). Earliest recollections of self and others in European, American and Taiwanese young adults. *Psychological Science*, 17, 708–714.

Warneken, F., & Tomasello, M. (2006). Altruistic helping in human infants and young chimpanzees. *Science*, 311(5765), 1301–1303.

White, B. L. (1985). *The First Three Years of Life.* Englewood Cliffs, NJ: Prentice Hall.

Wimmer, H., & Perner, J. (1983). Beliefs about beliefs: Representation and constraining function of wrong beliefs in young children's understanding of deception. *Cognition*, 13(1), 103–128.

Zheng, H. X., Yan, S., Qin, Z. D., & Jin, L. (2012). MtDNA analysis of global populations support that major population expansions began before Neolithic Time. *Scientific Reports*, 2, 745, doi:10.1038/srep00745.

5 Development of Climate Science

JULIA SLINGO

Introduction

'*Climate is what you expect, weather is what you get.*'[1] Therein lies an interesting question: what is the difference between weather and climate? It is of course just a matter of timescale; climate is, in effect, the statistics of the weather averaged over some time period, and, as I shall discuss in this chapter, the science of weather underpins the science of climate.

Climate Science is about understanding how the Earth's climate works at a global and regional scale; why it varies and changes through internal interactions, such as El Niño and the Thermohaline Circulation, and in response to external forcing agents, such as solar and volcanic activity; and whether human emissions of greenhouse gases will change, fundamentally, how the Earth's climate behaves. Not surprisingly, in recent years Climate Science has become synonymous with Climate Change Science.

As I will demonstrate in this chapter, Climate Science is about so much more than Climate Change. Climate Science, as a discipline, has emerged over my time as a scientist, from when I started as a researcher in the Met Office in 1972 following a degree in Physics. But its roots go back much further. It is synonymous with the disciplines of Meteorology, Oceanography, and Climatology, and it is rooted in classical Physics, Mathematics, Chemistry, and, increasingly, Biology. Modern Climate Science is fundamentally a fusion of theory, observations, and computational modelling.

[1] Frequently wrongly attributed to Mark Twain and more correctly linked to the geographer Andrew John Herbertson (1901) and to Robert Heinlein in his novel *Time Enough for Love* (1973).

To be a climate scientist requires one to be a polymath, a 'jack of all trades' if you will, and arguably 'master of none'. For that reason it can be viewed as not very rigorous, not 'proper' science, and not conducted in accord with the 'traditional' view of scientific endeavour, one of theory and testing by experimentation in a controlled environment. These arguments are often used to challenge the integrity and robustness of Climate Science and thereby to seek to undermine the significance of what it tells us about future risks to the Earth's climate from human activities.

In this chapter I will provide a personal perspective on how Climate Science has developed by highlighting a few key points in history, and by drawing on my own experiences over the last 40 years of how Climate Science has been transformed in that time through scientific and technological advances. Finally, I shall address the issues raised by human-induced climate change and how Climate Science can help us plan for a safe and sustainable future.

Historical Context

Climate Science has had a long and distinguished history. As a maritime and trading nation, England had a vast knowledge of the varying climates of the world, especially the winds over the oceans. In 1686 Edmund Halley published his iconic picture of the tropical winds in the *Philosophical Transactions* of the Royal Society – 'An Historical Account of the Trade Winds, and Monsoons, Observable in the Seas between and near the Tropicks, with an Attempt to Assign the Phisical Cause of the Said Wind' (Figure 5.1). Halley was curious why the winds invariably blew from the east, and argued that it must be due to the daily passage of the Sun, whereby the Sun heated the atmosphere, causing the air to rise and hence pulling air in from the east in the wake of the Sun's passage.

In 1735 it was George Hadley who postulated that in fact it is the Earth's rotation that drives the easterly trade winds. In a paper that was largely ignored at the time he wrote 'that the Air as it moves from the Tropicks towards the Equator, having a less Velocity than the Parts of the Earth it arrives at, will have a relative Motion contrary to that of the diurnal Motion of the Earth in those parts, which being combined with

FIGURE 5.1 Edmund Halley's map of the world's trade winds published in 1686 by the Royal Society. He depicted the major reversal of the trade winds between the winter and summer monsoons of Asia and Australia by using shorter dashes.

the Motion towards the Equator, a N.E. wind will be produc'd on this Side of the Equator, and a S.E. on the other'. He also realised that the greater heating from the Sun over the equator must cause the air to rise and that through continuity there must be an equivalent region of descent and the production of westerly winds away from the Tropics: 'The same Principle as necessarily extends to the Production of the West Trade-Winds without the Tropicks; the Air rarefied by the Heat of the Sun about the Equatorial Parts, being removed to make room for the Air from the Cooler Parts, must rise upwards from the Earth, and as it is a Fluid, will then spread itself abroad over the other Air, and so its Motion in the upper Regions must be to the N. and S. from the Equator.' From these ideas was born the Hadley Circulation, a fundamental part of the climate system.

It was not until a century later that Hadley's assertion that the Earth's rotation is fundamental really came to fruition. In 1835 Gaspard-Gustave de Coriolis introduced his theory of how objects move within a rotating frame of reference and the forces that act upon them. Coriolis didn't consider rotating spheres, but his theory was quickly taken up by meteorologists, including John Day, to explain the Earth's wind patterns. Hadley had been right in identifying the Earth's rotation as fundamental, but he had mistakenly assumed that absolute velocity was conserved rather than absolute angular momentum.

In 1856 William Ferrel provided the first explanation of the global circulation and the westerly winds, or passage winds as they were known then, that characterise mid-latitude climates. In his paper 'An essay on the winds and currents of the ocean' he wrote 'that when a particle of air receives a motion toward the poles it is deflected toward the east, as in the passage-winds, but when it receives a motion toward the south, there is a force which also turns it toward the west, as in the tradewinds. It has likewise been shown that when the air has a relative motion east, it has a tendency, on account of the greater centrifugal force, to move also towards the south, but that when it has a relative motion west, it has a tendency, on account of the diminished centrifugal force, to move also towards the north.' So by the end of the nineteenth century, through a combination of observations and theory, the fundamental importance of the Earth's rotation in defining the mean characteristics of atmospheric circulation, from the easterly trades to the mid-latitude westerlies, had been demonstrated.

The role of the Earth's rotation reached its ultimate expression in the work of Carl-Gustaf Rossby, who introduced the concept of absolute vorticity[2] and its conservation in adiabatic conditions – that is, without the influence of friction or heating from diabatic processes such as latent heat of condensation. He developed the theory of planetary waves – Rossby waves – within the atmosphere and oceans (Figure 5.2) and essentially laid the foundations of dynamical oceanography and meteorology.

In parallel to the development of our understanding of atmospheric circulation, physicists were trying to understand why the Earth has the temperature that it does, in other words its energy balance. Fourier had already shown that different gases absorb radiation in different spectral bands. John Tyndall used this information to prove that the Earth's atmosphere must have a Greenhouse Effect to explain its warm surface temperature; he also showed that these gases were emitters as well as absorbers of infrared radiation, which is vital for understanding the surface energy budget.

---

[2] Vorticity describes the local rotation or spin of a fluid – atmosphere or ocean. Absolute vorticity is the combination of spin from the planet's rotation and the relative spin of the fluid.

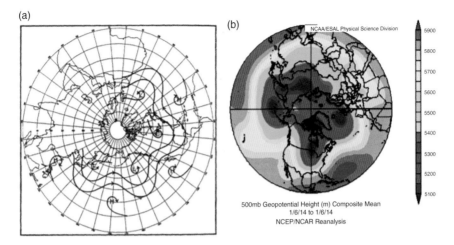

FIGURE 5.2 Rossby waves. (a) Rossby's example of waves in the mid-latitude westerlies (Rossby, 1940) and (b) a real example of Rossby waves in the mid-troposphere (500 hPa height field) on 6 January 2014 during the extreme cold event over North America (image provided by the NOAA-ESRL Physical Sciences Division, Boulder Colorado from their web site at www.esrl.noaa.gov/psd/).

Arrhenius took this a stage further in 1896 by making the first calculations of the influence of carbon dioxide on the Earth's surface temperature. In his book on *Worlds in the Making*, published in 1908, he states that 'any doubling of the percentage of carbon dioxide in the air would raise the temperature of the earth's surface by 4 degrees; and if the carbon dioxide were increased fourfold, the temperature would rise by 8 degrees'. This early climate-change projection, although at the high end of current projections, still lies within them.

It is interesting also to see how Arrhenius viewed human-induced climate change and its impacts:

> Since, now, warm ages have alternated with glacial periods, even after man appeared on the earth, we have to ask ourselves: Is it probable that we shall in the coming geological ages be visited by a new ice period that will drive us from our temperate countries into the hotter climates of Africa? There does not appear to be much ground for such an apprehension. The enormous combustion of coal by our industrial establishments suffices to increase the percentage of carbon dioxide in the air to a perceptible

degree ... By the influence of the increasing percentage of carbonic acid in the atmosphere, we may hope to enjoy ages with more equable and better climates, especially as regards the colder regions of the earth, ages when the earth will bring forth much more abundant crops than at present, for the benefit of rapidly propagating mankind.

Throughout the first half of the twentieth century concerns about climate change were very much focused on the possibility of entering another Ice Age based on what paleoclimate records from all sorts of geological evidence could tell us. In line with Arrhenius' thinking, global warming was not, as yet, a serious concern.

Returning to the dynamics of the climate system, there is another aspect of climate science which is of profound importance – that is, understanding how and why the climate of a region varies from year to year, and from decade to decade due to the internal variations in the climate system associated with oceanic and atmospheric flows and the interactions between them. If these can be understood, then it may be possible to predict variations in regional weather and climate patterns at least a season ahead.

While Tyndall and Ferrel were pondering the global aspects of the climate system, India was of growing importance for the economy of the British Empire. Indian cotton and grain harvests made up nearly one-fifth of the British economy and depended critically on the monsoon rains. Henry Blanford arrived in India in 1875 as the first British Director (Imperial Meteorological Reporter) of the Indian Meteorological Department. He found a climate where 'Order and regularity are as prominent characteristics of our [India's] atmospheric phenomena, as are caprice and uncertainty those of their European counterparts.'

But he was soon to be confronted by the great famine of 1876–1878, when the monsoon rains failed dramatically and the British economy was deeply affected. He decided that, because of the supposed simplicity of the Indian climate, it must be possible to find causes for these monsoon failures; through climate prediction (or seasonal foreshadowing as it was later called) famine could be anticipated and controlled, and India could be governed more effectively.

He and other meteorologists considered various possibilities, including the 11-year solar cycle. In 1884 Blanford published a paper in the *Proceedings of the Royal Society* entitled 'On the connexion of the Himalaya snowfall with dry winds and seasons of drought in India'. Commenting on 'the apparent dependence of the [1876] drought on the remarkable and unseasonable persistence of dry north-west winds down the whole of Western India', he noted that 'The experience of recent years affords many instances of an unusually heavy and especially a late fall of snow on the North-Western Himalaya being followed by a prolonged period of drought on the plains of North-Western and Western India.'

So began the growing body of research that sought to find relationships between the variations in climate in one region of the world and those in another, termed teleconnections by the British Meteorologist Sir Gilbert Walker.

In 1904 Sir Gilbert Walker arrived in India as the third British Director (Director General of Observatories) of the Indian Meteorological Department. He began to draw together observations from around the world, and pioneered statistical climate forecasting by constructing a 'human computer', with Indian staff performing a mass of statistical correlations using these data. As Walker said, 'I think that the relationships of world weather are so complex that our only chance of explaining them is to accumulate the facts empirically.'

From his endeavours came identification of the Southern Oscillation – and its association with failures of the Indian monsoon – the North Atlantic Oscillation, and the North Pacific Oscillation. In particular, the Southern Oscillation identified east–west variations in pressure around the equator, which describe changes in the preferred regions of ascending and descending motion – known as the Walker Circulation – and hence rainfall.

For the next 50 years, statistical climatology became a very important branch of Climate Science, through which empirical forecasting systems were developed to predict seasonal variations in the climate such as the Indian monsoon. But the causes of these climate variations were poorly understood; this is when oceanography enters the story.

The intermittent warming and cooling of the equatorial eastern Pacific Ocean – El Niño and La Niña – had been known for a long time, particularly by the Peruvian fisherman who saw their anchovy catch fail dramatically in El Niño years. In 1961, Vilhelm Bjerknes made the connection between this phenomenon in the ocean and the Southern Oscillation in the atmosphere, and the symbiotic relationship between the two – ENSO – was established. Although Henry Blanford did not know it then, the great Indian famine of 1876–1878 was caused by a very intense El Niño event.

So by the time I started my career in the Met Office in 1972, dynamical meteorology and weather forecasting, statistical climatology, paleoclimatology, and oceanography were well established and the transformation of Climate Science was about to begin.

Earth Observation: The First Transformation of Climate Science

We now know an immense amount about our climate, how it varies and changes, through a vast array of observations, especially from space-borne instruments. In the 1970s, what we knew was based primarily on the network of meteorological observations that were used in weather forecasting. These gave us a very limited view of the general circulation of the atmosphere and very little understanding of the role of the water cycle. At that time the first images from weather satellites were appearing, showing how clouds are organised, and by the early 1980s the first direct measurements of the Earth's radiation budget were being made. Over the subsequent decades the development of a constellation of satellites, geostationary and polar orbiting, has provided a rich resource for describing and monitoring the climate system (Figure 5.3).

We have been able to define the global flow of energy through the climate system with sufficient accuracy to know that the planet has been accumulating energy due to increasing atmospheric concentrations of greenhouse gases; and we know that around 90% of this additional energy is taken up by the oceans (Figure 5.4). We know that excess heat accumulated in the tropics is transported polewards, predominantly by the atmosphere in weather systems, and that phase changes of water, from evaporation (cooling) at the Earth's surface to condensation (heating) in the atmosphere as clouds and precipitation form, is a fundamental part of the Earth's energy cycle.

In fact, the ability of the Earth's climate to support water in its three phases – solid, liquid, and vapour – is one of the unique characteristics of

FIGURE 5.3 An example of the constellation of satellites that now observes many components of the Earth's climate system. This is complemented by a myriad of surface and *in situ* observing systems, including surface weather stations, weather balloons, aircraft, ocean buoys, floats, and ships. (Image courtesy of NASA.)

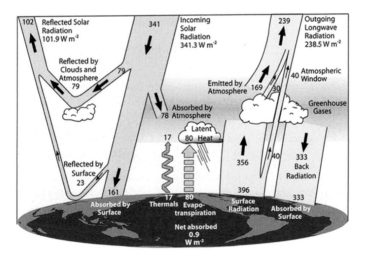

FIGURE 5.4 Flows of energy through the global climate system (W m^{-2}) from Trenberth *et al.* (2009). This emphasises that, although the balance at the top of the atmosphere is between the net shortwave radiation from the Sun and thermal, infrared radiation from the planet, at the Earth's surface the balance is much more complex. It involves other fluxes of energy besides radiation, predominantly from turbulent transports of moisture. In the atmosphere the balance is even more complex, involving clouds, emission and absorption of thermal radiation by greenhouse gases, and latent heat release. (© American Meteorological Society. Used with permission.)

(a)

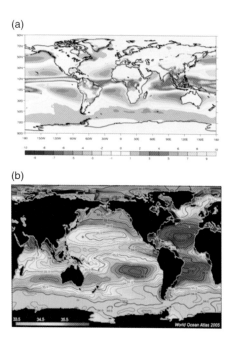

(b)

FIGURE 5.5 (a) Annual average climatologies of the difference between precipitation and evaporation (mm/day, where orange/red indicates an excess of precipitation over evaporation and blue indicates the opposite), and (b) sea-surface salinity.

the planet. It means that, between the Earth's surface and the troposphere, heat can be taken out in one location and released far way from its original source. This can be seen very clearly in the balance between evaporation and precipitation, where evaporation dominates in the subtropics and precipitation dominates in the deep tropics, near the equator (Figure 5.5). This means that much of the moisture released in the deep tropics has been transported by the atmospheric circulation, frequently the easterly trade winds, from far away across the subtropical oceans.

These regional differences between evaporation and precipitation imprint themselves on the salinity of the underlying ocean. Since salinity is a key determinant of ocean density, it is easy to see that the atmospheric water cycle is an important driver of the ocean circulation alongside the surface winds. In fact, a more detailed inspection of the atmospheric water budget reveals that the Atlantic is a source of moisture for the Pacific, explaining why the Atlantic is more saline than the other ocean basins, a feature that is important for the Atlantic Thermohaline Circulation and the Gulf Stream.

Today, Earth Observation tells us an immense amount about our climate system, its mean behaviour, how it varies, and how it is potentially changing. But it does not tell us why the climate system works in the way it does, how different components interact and drive the variability we observe, and why the climate might be changing. For that we need to use numerical models of the climate system.

Climate Models: The Second Transformation of Climate Science

In principle, fundamental physics tells us everything about the motion of the atmosphere and oceans, about the thermodynamics of the water cycle, about the transfer of radiation through the atmosphere, and about how the atmosphere interacts with the underlying surface. In practice we have to solve these physical equations on a computer by dividing the Earth's atmosphere and oceans into millions of volumes using sophisticated numerical techniques (Figure 5.6).

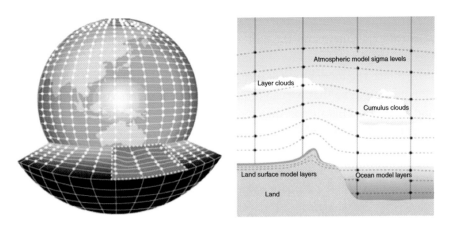

FIGURE 5.6 How a climate model divides the atmosphere, the oceans, and the surface into horizontal squares, typically 100 km or less in length, and how they are sliced up in the vertical. The fundamental equations of motion are solved numerically on this grid to simulate how atmospheric winds and ocean currents evolve over time.

The first climate models were known as general circulation models, and were developed in the 1950s alongside numerical weather prediction, which was also in its infancy. At that time the models were very simple in their construction, and the first simulations considered only the adiabatic flow without any representation of the hydrological cycle. It was shown very quickly that, in order to get anything like a realistic circulation, one required moist processes, but this raised some very big challenges that we still struggle with today.

The problem is that many of the processes that give rise to cumulus convection, condensation and the formation of clouds, and precipitation occur at scales much finer than those resolved by the grid of the model. Much of the early development of general circulation models therefore focused on finding ways to represent these sub-gridscale processes through the method known as parametrisation, in which the effect of these processes could be deduced from the resolved, large-scale characteristics of the atmosphere. This approach was remarkably successful, and enabled even very early models in the 1970s to capture fundamental aspects of the flow of energy around the climate system. Over the subsequent decades, these parametrisations have developed substantially, employing greater theoretical understanding, better observations, and the use of detailed laboratory and field experiments.

For climate scientists, the climate model is our laboratory. We cannot perform experiments on the real system to test hypotheses formed from theory and observations as one might in physics or chemistry. Instead we need to use the model to pick apart feedbacks and interactions within the climate system so that we can understand how it works and why it varies and changes. This means that we are always testing the validity of our models against theory and observations, and always seeking to improve their skill.

Over the last few decades models have enabled us to understand so many aspects of the climate system, from how soil moisture feedbacks affect the West African Monsoon, via how what is happening in the tropical West Pacific drives the climate over North America, to how the 11-year solar cycle affects the winter weather we experience in the UK. We have learnt how mountains affect the position of the storm-tracks, whether Himalayan snow cover can really influence the progress

(a)

(b)

(c)

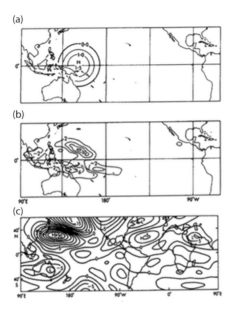

FIGURE 5.7 An example of the use of a climate model to understand the possible relationship between some 'severe' winters in North America and enhanced convective activity over the tropical West Pacific (Palmer and Owen, 1986). (a) Impose sea-surface temperature anomaly. (b) The model simulates the local rainfall (heating) response. (c) The model develops the global changes in the atmospheric circulation through the excitation of Rossby-wave trains. In this case, the circulation changes over the North Pacific and North America drive a cold north-easterly flow over North America similar to the severe winter of 2013/14 when rainfall was much higher than normal in the tropical West Pacific. (© American Meteorological Society. Used with permission.)

of the Indian Monsoon, how the pattern of ocean temperatures in the North Atlantic affects European climate, and many more important findings. All this has been achieved through careful design of 'what if' experiments based on hypotheses drawn from observations of the past and present climate (Figure 5.7).

The climate models of today are now effectively simulators of the real world. They are able to produce realistic simulations of the weather, monsoons, El Niño and its impacts, the Atlantic Thermohaline Circulation, and the Gulf Stream to name but a few phenomena. These properties of the simulated climate system emerge as a result of only two fundamental constraints – the energy input from the Sun and the rotation rate of the planet. Even the composition of the atmosphere is determined from the moist processes, emissions of greenhouse gases and aerosols, and chemical reactions. The construction of these models is arguably one of the great scientific achievements of the last 50 years.

Supercomputing: The Third Transformation of Climate Science

Ever since their inception, climate models have been very computation-intensive, and the availability of computing power has dictated the level of sophistication and the type of experiments that can therefore be performed (Figure 5.8). There are few sciences where progress can be so closely linked to the increases in supercomputing power.

Supercomputing has transformed Climate Science. It has enabled increases in resolution so that the models capture more faithfully the weather systems and ocean eddies that constitute the climate (Figure 5.9);

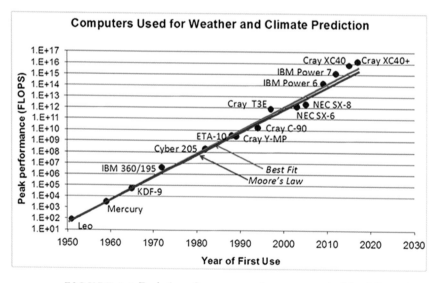

FIGURE 5.8 Evolution of supercomputing power at the Met Office, which has contributed to substantial improvements in weather forecasting and climate prediction.

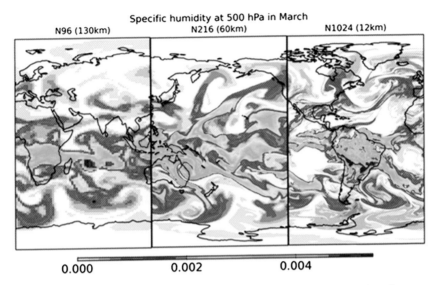

FIGURE 5.9 An example of the improvements in the description of atmospheric flows with higher model resolution. This shows the moisture field in the middle troposphere and highlights the importance of resolution for capturing the filamentary structures that are known to be to be critical for the water cycle and the formation of precipitation along weather fronts and in tropical convective systems. (Image courtesy of the Met Office.)

it has allowed the introduction of more components of the climate system and the transformation to Earth system models; and it has delivered the capability to perform a large number of simulations to test for robustness.

Increasing the horizontal resolution is particularly challenging; halving the grid length leads to a requirement for 10 times the computing power, so it is not surprising that climate models have lagged behind weather forecast models in terms of their ability to capture weather systems. Even the latest models used in the IPCC Fifth Assessment Report (2013) had resolutions coarser than 100 km. However, that is changing rapidly now as more computing power is becoming available and there is greater appreciation of the scientific need to resolve atmospheric motions – the weather – and how they transport heat, momentum, moisture, and other atmospheric constituents.

The ocean is potentially even more challenging, because the scale of the eddies (the equivalent of weather systems) is a fraction of those in the

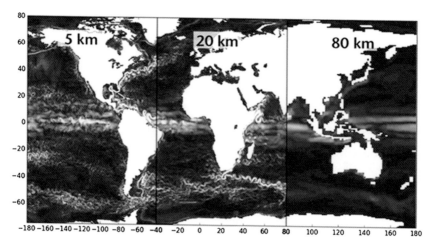

FIGURE 5.10 An example of the improvements in the description of upper-ocean flows with higher model resolution. This shows the sea-surface height and highlights the importance of resolution for capturing ocean eddies and western boundary currents, such as the Gulf Stream. (Image courtesy of the Met Office.)

atmosphere (Figure 5.10). The resolutions of order 80 km typically used in climate models require the effect of ocean eddies to be parametrised and compromise important components of the ocean circulation, such as the Gulf Stream. The latest climate models, which use significantly higher ocean resolution of order 20 km, can begin to capture ocean eddies and are leading to substantial increases in skill, but it is generally believed that resolutions of order 5 km are necessary to represent ocean eddies. As well as being important for transporting heat around the ocean, regions of eddy activity are also those where biological production is high, so they are also critical for the uptake of carbon by the ocean.

Over time, climate models have evolved from atmospheric models to considering many other components of the system, namely oceans, sea-ice, the land surface, the carbon cycle, chemical composition, and so on. They are now effectively models of the Earth system (Figure 5.11), and arguably represent some of the most complex codes ever written.

But complexity comes at a computational cost, and so the resolution of these models was compromised to enable different interactions and feedbacks within the climate system to be represented. It is now increasingly recognised that many of these interactions and feedbacks operate on

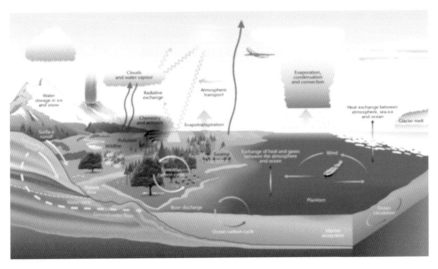

FIGURE 5.11 A schematic representation of the processes represented in current Earth system models. (Image courtesy of the Met Office.)

space and time scales that relate to atmospheric weather and ocean eddies, and recent advances in supercomputing power are allowing simulation of these phenomena.

Uncertainty is an inherent property of the fluid motions of the atmosphere and oceans, which determine the weather and climate at the regional and local level. This was recognised in 1963 by Ed Lorenz in his seminal paper on 'Deterministic non-periodic flow' in which he introduces the concept of the atmosphere as a chaotic system subject to small perturbations that grow through non-linear processes to influence the larger scale – 'the flap of a seagull's wings may forever change the course of the weather'.

The concept of the weather and climate as chaotic systems has had a profound impact on the way in which modelling and prediction have evolved over recent decades. No longer do we produce a single prediction, but instead we perform an ensemble of predictions that seek to capture the plausible range of future states of the weather and climate that might arise naturally from 'the flap of the seagull's wings'. Again this places demands on the computing power available to us.

So the climate scientist is always making choices about how best to deploy available supercomputing resources, whether to trade off resolution for complexity or ensemble size. There is never a single answer; it depends on the scientific application and on our level of understanding of what those choices mean for the validity of the simulation or prediction. There is no doubt, however, that the inadequate availability of supercomputing power continues to hold back Climate Science and that a compelling case can be made for greater investment.

Global Warming: The Fourth Transformation of Climate Science

In 1958 Charles David Keeling began making measurements of atmospheric concentrations of carbon dioxide (CO_2) at Mauna Loa, and soon began to notice that the concentrations were rising systematically year-on-year. And so the huge influence that human-induced climate change would have on climate science began.

The first simulations of the possible implications of increasing CO_2 were performed in the 1970s, and by the early 1980s were an integral part of climate research in the Met Office. The need to understand the sensitivity of the climate system to greenhouse gases undoubtedly had a big influence on model development. From the introduction of a fully interactive ocean model to address ocean heat uptake, via the development of terrestrial vegetation and ocean biogeochemistry models to understand the role of the carbon cycle in amplifying global warming, to complex cloud microphysics to understand cloud feedbacks, and interactive sea-ice models to address polar amplification, these have been major endeavours involving national and international science partnerships.

It would be wrong to conclude, however, that Climate Science has been taken over by the politics of climate change as some would claim. Beyond climate change there has been an immense body of research on natural climate variability from which has come the capability to predict a season, or even longer ahead. Remembering that natural climate variability is still the dominant cause of climate extremes around the world, the value of these advances is considerable. Nevertheless, climate change has presented some significant science challenges that would not have happened otherwise, and therefore has accelerated the progress of Climate Science over the last two or three decades.

By the time the IPCC published its Fifth Assessment Report in 2013 the evidence for a warming world was unequivocal, with rising surface temperatures, melting ice sheets, and rising sea levels. The IPCC further stated that it is 'extremely likely (95–100%) that most of the observed increase in global surface temperature since 1951 is caused by human influence'. That statement was based on the use of climate models to investigate what the world's climate would have been like without human emissions of greenhouse gases and land-use change. Without the development of sophisticated climate models in the last 50 years the attribution of global warming to human factors could not have been made.

Attribution of climate change now goes beyond just considering the global mean surface temperature to address other components, more regional aspects of the climate system, and even extreme events such as flooding, drought, and heatwaves (Figure 5.12). Year-on-year there is

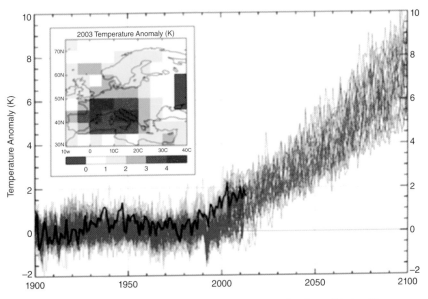

FIGURE 5.12 A Met Office study of the attribution of higher European summer temperatures to climate change. In 2003 Europe experienced extreme heat that was formally attributed to climate change (Stott *et al.*, 2004). Since then observed temperatures have continued to be high (black line) and now fall well outside the envelope of model simulations (blue) without anthropogenic forcings. Instead they fall within the upper range of the simulations with observed concentrations of greenhouse gases. (Image courtesy of the Met Office.)

ever growing evidence that human-induced climate change has made a contribution to the severity of these sorts of event. Furthermore, climate change can now be linked to the impacts of severe weather, such as coastal inundation from Hurricane Sandy, where rising sea levels due to climate change were a contributor.

Despite all the debates about uncertainties in climate models and in the projections of climate change, arguably one of the most important figures from the IPCC Fifth Assessment Report was the very simple and fundamental result that if we continue to accumulate carbon in the atmosphere then the planet will continue to warm, just as Arrhenius hypothesised in 1896. In fact the relationship is almost linear (Figure 5.13), and makes

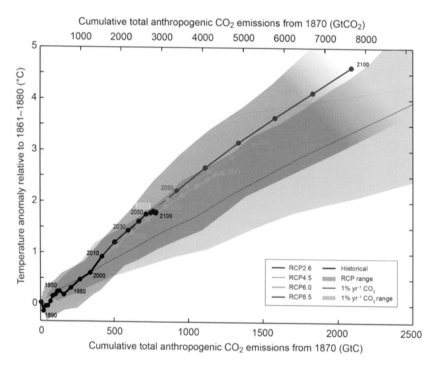

FIGURE 5.13 The relationship between the accumulation of carbon emitted from human activities since 1850 and the change in decadal average global mean surface temperature, from observations up to 2011 and projected by models based on a range of scenarios of annual CO_2 emissions to the end of the century. To limit warming to 2 °C implies that total emissions should be limited to 1,000 gigatonnes of carbon ('trillionth tonne'). (From IPCC AR5.)

clear what the choices we make around reducing emissions will mean for global warming.

The sophistication of the climate models used to predict global warming means that we can say a lot more about what climate change will be like at the regional level and for many aspects of the climate system beyond temperature. We know that parts of the world that are already water-stressed will become more stressed, that the Arctic is likely to be ice-free in summer within the next two or three decades, that the incidence of heat stress will increase, and that extreme rain events will become more severe. Since CO_2 can remain in the atmosphere for centuries, we know that what we have emitted so far will continue to affect the climate for decades to come, and even after global temperatures have stabilised sea levels will continue to rise.

Mechanisms of Change: Helping Us Plan for a Safe and Sustainable Future

In 1990, at the time of the publication of the first IPCC Report, Prime Minister Margaret Thatcher had the foresight to establish the Hadley Centre for Climate Prediction and Research in the Met Office. Building on the modelling expertise developed for weather forecasting, this enabled the UK to become the world-leader in Climate Science that it is today, and to make substantial contributions to subsequent IPCC reports. Her words from her opening speech at the opening of the Hadley Centre on 25 May 1990 are as relevant today as they were 25 years ago:

> We can now say that we have the Surveyor's Report and it shows that there are faults and that the repair work needs to start without delay . . . We would be taking a great risk with future generations if, having received this early warning, we did nothing about it or just took the attitude: 'Well! It will see me out!' . . . The problems do not lie in the future – they are here and now – and it is our children and grandchildren, who are already growing up, who will be affected.

The evolution of Climate Science means that today it is ready to play a central role in helping us plan for a safe and sustainable future.

The predictive power of climate models enables us 'to see into the future' based on fundamental physics, so that we can be better prepared to deal with the risks we face from human interference with the climate system. The scenarios of the future that climate models provide us with truly act as mechanisms of change; they provide a framework for action and enable us to provide answers to the following questions.

- Can we provide society with a 'road map' indicating what climate variations and changes may be expected to occur, where, and with what implications?
- How can we make society more resilient and better prepared for hazardous weather and climate extremes arising from climate variability and change?
- Do we know what levels of climate change could be dangerous, where, and for whom?
- What should society do to mitigate and adapt to climate change to avoid its worst impacts?

The transformation of Climate Science is reflected in the work of the IPCC. Nowhere else in science do so many scientists come together to provide the world with the evidence it needs to take the actions that will determine the future of life on our planet. Working Group 1 – The Physical Science Basis – from the Fifth Assessment Report involved 859 scientists who gave their time freely as authors and editors; they reviewed and summarised 9,200 papers, they engaged 1,089 expert reviewers, and they responded to 54,677 comments. The IPCC is one of the great services of science to society.

Finally, it is worth returning to where this chapter began – 'Climate is what you expect, weather is what you get.' Through the development of Climate Science outlined here, we have begun to appreciate more and more that there is no distinction between the weather and the climate; the same science underpins them both. As we look to the future and a warming world, the greatest impacts of climate change on society will be felt through changes in the weather, especially hazardous weather such as floods, storms, and heatwaves. The transformation of Climate Science to a science that is rooted in local weather is the next big step.

References

Palmer, T. N., & Owen, J. A. (1986). A possible relationship between some 'severe' winters in North America and enhanced convective activity over the tropical west Pacific. *Monthly Weather Review*, 114, 648–651.

Rossby, C.-G. (1940). Planetary flow patterns in the atmosphere. *Quarterly Journal of the Royal Meteorological Society*, 66, Supplement, 68–87.

Stott, P. A., Stone, D. A., & Allen, M. R. (2004). Human contribution to the European heatwave of 2003. *Nature*, 432, 610–614.

Trenberth, K. E., Fasullo, J. T., & Kiehl, J. (2009). Earth's global energy budget. *Bulletin of the American Meteorological Society*, 90, 311–323.

6 Biomimicry: Development of Sustainable Design

MICHAEL PAWLYN

Introduction

Biomimicry is a subject that is going to be one of the most important sources of new solutions in the decades ahead, possibly the most difficult decades that humanity will have faced.

One of my favourite quotations is from the musician John Cage, who said 'Some people are scared of new ideas; I am scared of old ones.' In this essay, I will discuss the scary old ideas as little as necessary and the new ideas as much as possible, because I want to focus on solutions. I will explain what I mean by biomimicry (and give a few examples of that), but first of all I will give some context and explain how this can influence a rather new direction in architecture.

It is a very difficult time for those who are committed to the environment; many of the world's governments are unfortunately asleep at the wheel, and there is also a very unhelpful chasm between conventional economists and environmentalists. Many of the former argue that there is not a strong enough case for action, and many of the latter use such gloomy language that it is counter-productive – it is disempowering. We are increasingly accustomed to seeing graphs showing carbon dioxide levels and temperature over time. Many of the trends that we are seeing in the environment are exponential and yet still the economists do not seem to accept the case for action.

There is a very nice little anecdote about when Satish Kumar went to visit the head of the London School of Economics and his opening gambit was 'Do you have a department of ecology?' The chap looked a bit bemused and said 'Err ... no.' And Satish then said 'Do you know the derivation of the word economy?' 'Err ... not sure.' 'Well it comes from the Greek word *oikos* meaning "home" and *nomos* meaning "management"; which is very similar to "ecology" which comes from *oikos* meaning "home" and *logos* meaning "knowledge". So, how can you manage your home without knowledge of it? You are producing students who are only half educated!' I do not think Satish was invited back after that, but he made an important point. There are still some sceptics who claim that we really do not have a problem, while at the same time there are increasingly alarming analyses coming from scientists, predicting that we could face a perfect storm of resource shortages, climate chaos, and over-population.

We urgently need a more sophisticated approach to economics. When looked at from an ecological perspective, conventional economics is woefully inadequate. We need approaches that accommodate complexity.

Biomimicry

To begin, I would like to explain what I mean by 'biomimicry', by giving a couple of examples.

There is a beetle called the 'bark beetle', which can detect a forest fire from somewhere between 10 and 80 kilometres depending on which scientist you believe. If one compares that with conventional man-made fire detectors, which have a range of about 8 metres, the beetle is about 10,000 times as sensitive. Moreover, it does not need a continuous connection back to a power station burning fossil fuels. It is an amazing adaptation to a very specific niche. Who would have thought such sensitive fire detectors would exist in biology? If we could learn how it does that then we could start developing much more effective, much more energy-efficient, fire detectors ourselves.

Another example is the 'bombardier beetle', which defends itself against predators by firing a high temperature explosion out of its abdomen (Figure 6.1(a)). It has a combustion chamber in which it mixes two highly unstable chemicals, hydroquinone and hydrogen peroxide. To stop the combustion migrating back into the fuel tanks, there are inlet valves that open and close 200 times a second. Engineers have been studying this beetle in order to develop more efficient fuel-injection systems, needle-free medical injections, and even a new form of fire extinguisher. The closest humans have come to this is the German Messerschmitt Me 163 rocket airplane

(a)

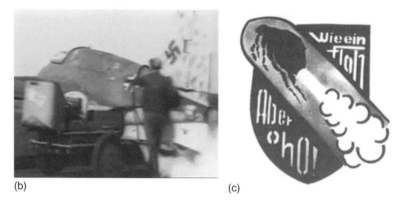
(b)   (c)

FIGURE 6.1 Biomimicry: the bombardier beetle and the rocket plane.
(a) A bombardier beetle ejecting an explosive mix of hydroquinone and hydrogen peroxide
(Eisner & Aneshansley, 1999. Copyright (1999) National Academy of Sciences, USA).
(b) Footage from the 1940s of a German Messerschmitt Me 163 rocket airplane and (c)
the insignia used by one squadron of Me 163 pilots.

(Figure 6.1(b)), which mixed together exactly the same compounds, hydroquinone and hydrogen peroxide. The German designers back in the 1940s
did not have fuel-mixing technology as efficient as what we have nowadays,
so a lot of these aircraft burst into fireballs on the ground. However, the
insignia used on the side of the plane by one squadron (Figure 6.1(c) – the
text 'Wie ein Floh, aber oho!' translates roughly as 'Like a flea – but oh gee!')
suggests that they may have been aware of the beetle.

My favourite example of biomimicry is the 'dog vomit slime mould'
(Figure 6.2(a)). We need more of this, particularly in cities. Slime
moulds are single-celled organisms that form minimum-distance networks between sources of food. In 2009, scientists at Hokkaido

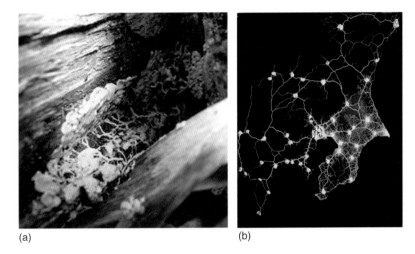

(a)　　　　　　　　　　　　　　　　　(b)

FIGURE 6.2 Biomimicry: the 'dog vomit slime mould'. (a) An example of a slime mould living in the wild. (© Bernard Bradley, Creative Commons.) (b) Hokkaido University experiment leading a slime mould to expand to mimic the railroad network of the greater Tokyo area. (© Mark Fricker, Atsushi Tero, Seiji Takagi, Tetsu Saigusa, Kentaro Ito, Dan P. Bebber, Mark D. Fricker, Kenji Yumiki, Ryo Kobayashi, and Toshiyuki Nakagaki.)

University carried out an experiment in which they took a map of the Tokyo region and they put a small source of food on each of the cities surrounding Tokyo and a slime mould on Tokyo itself. The slime mould spread out quite quickly, it located all those sources of food, and then it started optimising the connections between them. When that slime mould had finished, the layout exactly matched the railway network in that part of Japan (Figure 6.2(b)). It had taken the engineers thousands of hours to arrive at that optimisation, whereas the slime mould did it in 26 hours, which suggests that we do have a bit to learn. The key point is that, if we could learn from biological organisms to develop biologically based algorithms, then we could start to create much more efficient networks and design much more efficient buildings. Overall, I think you could look at the living natural world as being like a design sourcebook in which all the products have benefited from a 3.8 billion year research and development period. Given that level of investment, it makes sense to use it. We are looking at the results of a long and ruthless process of evolution, which has left us with some amazing success stories that we can learn from.

I think there are three really big challenges we need to address over the next few decades. The first one is achieving radical increases in resource efficiency, so doing more with far less. The second one is shifting from linear, wasteful, and polluting ways of using resources to closed-loop models, in which all the resources are stewarded within closed loops, and no waste is lost and no pollution is emitted. The third one is the shift from a fossil fuel economy to a solar economy. If we choose to embark on these three interlinked journeys then, in my opinion, there is no better source of ideas than biomimicry.

In what follows, I will describe some projects we have worked on that have used these ideas. That will perhaps convey a bit more clearly what biomimicry is and how it can be used.

The Biomimetic Office Building

This project was our opportunity to use biomimicry to comprehensively rethink office design. We persuaded the client to allow us to do a concept study first with a minimum of constraints (so no particular site, a very loose brief) and to appoint a really excellent team. Perhaps I am being rude about my profession, but the way in which architects towards the 'iconic' end of the spectrum work with their consultants is that sometimes they will just do the seductive sketch and then expect the rest of the team to just make it happen. If you have gone to the trouble of appointing a really excellent team then that approach is verging on insulting. I think you need a much more collaborative approach to draw the best ideas out of the team. For me, the model is really a conductor. A conductor does not necessarily make any sound at all, but relies for his or her power on making other people powerful, and drawing that together in a unified vision. That is the kind of thing that inspires us, and this was a wonderfully enjoyable exercise, working with a fantastic team, including a biologist. In the first workshop we concluded that we wanted to achieve a minimum 10% increase in productivity for the occupants, for the building as far as possible to be passively self-heating and cooling, to be entirely lit with daylight, to be a net producer of energy, and for the air coming out to be cleaner than the air going in. The last three points, I would argue, are going beyond sustainable design to achieve restorative, or regenerative, results. I do think it is time we move beyond the sustainability paradigm, which all too often was about

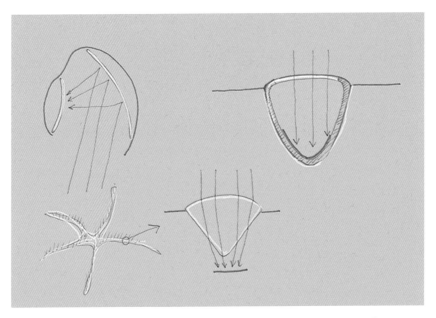

FIGURE 6.3 Light gathering and distribution in biology. Sketches illustrating light focusing by spookfish eyes (top left), stone plant leaves (top right), and the brittlestar (bottom). (© Exploration Architecture.)

mitigating negatives, to a much more positive mind-set which is about optimising positives and achieving regenerative results.

The team decided that one of the most powerful drivers of the building form was likely to be daylight and, with our biologist, we set about studying ways in which light is gathered and distributed in biology. We looked at organisms such as the spookfish, which have amazing mirror-shaped eyes, and these mirror structures are able to focus low-level bioluminescence coming up from the deep ocean onto the fish's retina (Figure 6.3, top left).

Another one we looked at was the stone plant. These are plants that live in deserts and for reasons of thermal stabilisation most of the plant is below the ground, so that the chemical reactions can take place at a fairly steady temperature. All the chlorophyll is down in what you could call the basement and it has a kind of roof light in the top that allows the light down to where the chlorophyll is (Figure 6.3, top right).

A further example we studied was the brittlestar. This is a starfish that lives as much as 500 metres below the ocean surface, and it has evolved a covering of nearly optically perfect lenses over its skin that are able to detect very small amounts of light and movement and focus them onto receptors so that it can see predators long before they can see it (Figure 6.3, bottom).

These examples encouraged us to think much more creatively and deliberately about how we would bring light into this building. A conventional way of designing for daylight in offices is to just think about the right distances between the windows. In London you can often find really deep offices that are 25, or even 40, metres deep, and you know that those buildings are going to be very energy-intensive because they are dependent on artificial lighting and air conditioning. We concluded that the right dimension was about 12 metres, so no one is further than 6 metres from the nearest window. Then, thinking about what kind of building form that suggested, one approach would be to stack the narrow floor-plates up into a tall tower, which is fine if you are working in a dense urban location with high land values, but we wanted to create a concept that was more universal than that. We looked at two other building forms: a ring of offices around a central atrium and a more linear approach with two linear blocks and an atrium down the middle. It was the third one that seemed to work best when we analysed the light levels. What we found from the daylight modelling was a curved pattern of shading towards the middle of each block because of the shading effect of the opposite block. So the next move was to simply bend those floor-plates to produce an even quality of light all the way along.

This produced two further challenges. The first is that narrow floor-plates are not particularly good for creative clusters of people; you need some parts of the floors that are wider, and this was not making very efficient use of the rectangular site. Learning further ideas from biology of surface-area-to-volume optimisation, we elaborated the plan into this undulating form, so now we have much better facilities for creative clusters of people and still no one is further than about 6 metres from the nearest window (Figure 6.4, top panel).

Still looking at daylight, we found that we could fairly easily get enough light in the upper parts of the building – the second challenge was how to get light further down. So we looked at the possibility of harvesting light near the top and focusing it into fibre-optic tubes and channelling it around the building. For this we looked at a rainforest plant called *Anthurium warocqueanum*, which lives in very-low-light conditions and has lenses all

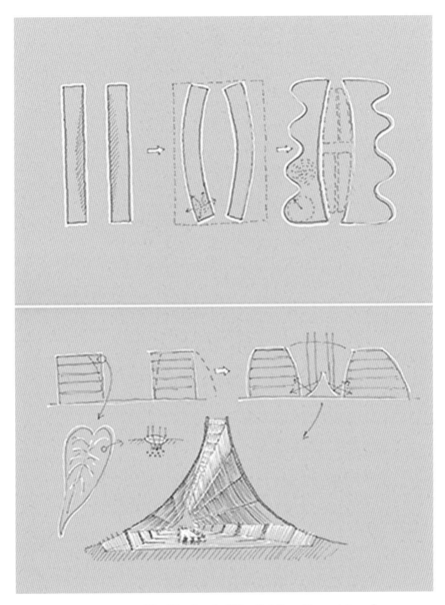

FIGURE 6.4 The Biomimetic Office Building. Floor-plate and
section sketches. (© Exploration Architecture.)

over its leaves. This proved to be pretty complicated and there was not the time to develop that within the constraints of this study, so that has become a separate research project. What we did conclude was that it was worth shaping the building to incorporate a large-scale light reflector, like the spookfish eye, that would bounce light into deeper parts of the building. Thinking about what we would do underneath the reflector, this was a great opportunity to create a really dramatic meeting space that would add value to the building and would have some of the qualities of the film-sets Ken Adam designed for Stanley Kubrick and for James Bond films (Figure 6.4, bottom panel).

Daylight was really the primary driver of the architectural form so, when it came to designing the structure, the way we used biomimicry was to minimise resource use as much as possible. We wanted to use material much more efficiently and so we started looking at biological structures like bird skulls. Biology often reveals very complex structures that achieve their resource-efficiency by putting the material exactly where it needs to be. In the case of bird's skulls you have these very thin layers of bone connected together with struts and ties, like a combination of dome technology and space-frame technology. We also looked at cuttle-bone, which is quite similar – again you have very thin layers of bone with undulating connect-ing walls, creating a very stiff structure with an absolute minimum of resources. Next, we analysed a generic floor slab and column arrangement in terms of which parts were working hard structurally and which bits were rather redundant. What we found is that much of the volume is structurally redundant, namely the centre of the columns, the middle of the floors, and so on. If we were to shape the structure so that it followed the shape it wants to be, it would probably end up with hollow columns because the centre is not doing anything structurally, there would be slightly deeper floors where the bending moments are highest, and we could take out some of that mass with void formers. These are technologies that already exist, like fabric formwork and bubble-deck – we're just pushing them a little bit further. This was getting reasonably close to the examples we started with, and this demonstrates why biomimicry can be a very powerful tool for designers and engineers. It allows the conversation to wander off in pursuit of the most ideal solutions, and once you have established that, then you can come back to something that is achievable within the constraints that you are working with. I take the view that you should never start with reality, you should always start by identifying the ideal and then compromise as little as necessary.

We also learned from many more examples: from termites to design a passive heating and cooling system; from plants like *Mimosa pudica* and from folding beetle wings to design a solar shading system which lets in exactly the right amount of light and converts all surplus light into electricity; and from curved leaf shapes and shell forms to design a new glazing system that is based on very thin curved pieces of glass, which should achieve a roughly 50% saving in glass on the building.

Figure 6.5 shows how the scheme looked in 2015. The image in part (a) shows the strategic form driven by daylight, the spookfish light reflector in the atrium, and the curved glass units. The image in part (b) shows the atrium with the spookfish light reflector meeting room in the middle. You can see that, instead of creating gloomy gas-guzzling buildings that are soul-sapping to the occupants, it is perfectly possible to create really uplifting, light-filled spaces that are inspiring for people to work in and regenerative to the human spirit.

This was only a few per cent more expensive than a more conventional approach. We did a back-of-the-envelope calculation for the client and worked out that the building cost was about 90,000,000 Swiss Francs and the annual salary cost for the people inside would be about 300,000,000 Swiss Francs. If you can achieve a 10% increase in productivity (which is

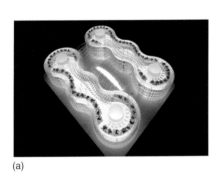
(a)

(b)

FIGURE 6.5 The Biomimetic Office Building. (a) CGI of the strategic form driven by daylight and (b) the spookfish light reflector meeting room. (© Exploration Architecture.)

perfectly possible – there are peer-reviewed experiments which show a 20% increase), that would be worth 30,000,000 Swiss Francs per year, which means that the building would pay for itself completely in 3 years. It is quite difficult to convince clients (and certainly conventional econo-mists) of this approach, but we have seen a shift in thinking. Ten years ago all people seemed to be talking about was the value of energy saved, and actually that is not a big sum of money, particularly when talking about a commercial office building. Now people are talking much more about the importance of looking after your staff, making sure they are happy and productive. Companies like Google spend enormous amounts just to attract the right people, and they will justify expenditure on things that sometimes have very long payback periods. They do it because it raises the 'coolness factor' of the company so that they attract the right people and they see all sorts of benefits. That is the kind of thing I mean when I refer to a more comprehensive view of economics.

The Mountain Data Centre

Office buildings have been improving and have been getting steadily more and more energy-efficient over the last few decades. The one area that has been getting steadily worse is IT-related energy consumption, and that is partly because of the way we use data centres and servers. Often they are in buildings or urban areas and they need a huge amount of cooling, which substantially increases energy consumption. So what could biology possibly teach us about data centres? They do not exist in biology, and this is where it is important to remember that biomimicry is a functional discipline. It is about identifying how functions have been delivered in biology and then translating that understanding into solu-tions that suit human needs. If you study how animals keep themselves cool (they do not have access to intense forms of energy like fossil fuels), you can observe that they just do simple things, like moving somewhere that is cooler or engaging in evaporative cooling by panting or sweating.

The first move on this project was to locate the server somewhere that is already extremely cold and rely on high-speed data transmission to get the data to the users. The cold location in this case is a mountain in Norway that has already been carved out for marble mining, and there are about 90 linear kilometres of tunnels that are all at a temperature of about 5 °C all year round. So then the challenge was 'how do we draw that really cool air through the data centre to keep it cool?' This time we looked at a phenomenon in biology referred to as 'Murray's law'. This is a mathematical principle that seems to be true of most branching systems in biology. There is a cubed ratio between the vessels in branching systems, so the cube of the diameter of the parent vessel is equal to the sum of the cubes of the two daughter vessels. There also seems to be a very consistent angle, which follows principles of hydrodynamics and aerodynamics, so overall this appears to be an evolved minimum-energy solution. Conventionally data centres are laid out in a very linear way, with data blocks in lines with long runs of ductwork and a lot of bends. We proposed to arrange the data blocks (the individual components of the data centre) into a circular cluster, and at a stroke that reduced the length of ductwork enormously and the number of bends. We then used Murray's law to optimise that branching network in order to arrive at a theoretically optimised extract system. So this would draw the cool air through the data centres and keep them functioning using a minimum of energy. The showpiece data centre in the centre of the mountain does have a bit of a James Bond feel about it. First of all you have to come through a scanner and you go down a long tunnel into the mountain. Then you get into a boat to cross an underground lake and you end up at the cavern in the centre where our showpiece server is. Then there is a shaft that will take the exhaust air up to a disused quarry at the top where the client says he wants to do a mini-Eden Project.

At the beginning I mentioned three big transformations: radical increases in resource efficiency, going from linear to closed loops, and the passage from fossil fuel to a solar economy. So the two projects we have looked at so far were mainly about achieving radical increases in resource efficiency, designing an office building that was far more energy-efficient, and producing a data centre that used radically less energy.

The Cardboard to Caviar Project

Next I want to turn to a project that explores ideas of closed-loop design. Conventionally humans have tended to use resources in a linear way: we dig things out of the ground, we use them, and we dispose of them. A statistic from WRAP (Waste Resource Action Programme) indicated

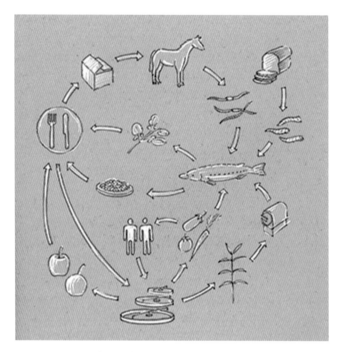

FIGURE 6.6 The Cardboard to Caviar Project. (© Exploration Architecture.)

that 70% of what we take out of the ground is back in the ground within a year, which is phenomenally wasteful. Mature ecosystems work very differently from this: all the resources are stewarded within closed-loop cycles, the nitrogen and the carbon and so on. There have been quite a few industrial projects that have attempted to mimic this idea, modelling themselves on ecosystems. There is one in Denmark called 'Kalundborg' in which a number of industries were co-located, so they could all benefit from each other's waste streams. The waste from one becomes the input for something else in that system. The individual components of Kalundborg were not particularly benign, there was a fossil fuel power station, chemical works, and so on.

Another project, the Cardboard to Caviar Project, was set up by Graham Wiles in Kirklees in Calderdale (Figure 6.6). They collected cardboard from shops and restaurants and then they shredded it and sold it to equestrian centres as horse bedding. When that was soiled, they were paid to collect the manure and cardboard. They then put it into wormery compost systems which produced worms, which they fed to Siberian sturgeon, which produced caviar, which they sold back to the restaurant. They turned a linear system into a closed-loop system and they actually earnt money along the way: they were paid for the bedding, they were paid to take away the soiled bedding, and they were paid for the fish and the caviar.

That is the overview, but in some ways the fuller picture is much more interesting. When they started, it was just the first part of this diagram. Graham was involving people with disabilities in a recycling initiative, shredding cardboard and selling it to an equestrian centre, and that was as far as it went. And then the horsey folks said 'Well, what do we do with all this soiled cardboard and manure?' and Graham's first idea was to set up a wormery and to sell the worms to a fishing bait supplier. At the eleventh hour the fishing bait supplier backed out of the deal, so Graham decided to cut out the middleman and set up his own fish farm. This time he started working with reforming heroin addicts, who were quite a difficult bunch of teenage kids on the whole, and they were costing the local authority as much as £100,000 per addict per year on rehabilitation schemes with a 95% failure rate. Graham has achieved a 65% success rate at getting the kids off drugs and into something more productive with a fraction of the amount of money.

They found that the fish were not putting on enough weight in the winter because the water was too cold. By this stage they had been given quite a big area of former industrial land that was next to a Yorkshire Water treatment facility. They used some of the treated fertiliser sludge from the sewage works to restore this land and planted willow biomass, which would run a biomass boiler to keep the fish happy in winter. Then they found that they were spending a certain amount of money on fish food to supplement the worms, and Graham did not like anything leaking out of the system, so he got the kids growing vegetables in order to make their own fish food. As it turned out, the kids ate all the vegetables, but in many ways that was a good thing because prior to that they were coming to the site with a can of Coke and a Mars bar for their lunch, but now they were learning about growing and eating healthier food. Then they restored more industrial land, they planted orchards (creating another product that could be sold back to the shops and restaurants), they redesigned the water treatment for the fish tanks, and they created these long planting beds for watercress that would take out all the excess nutrients and produce another foodstuff for the shops and restaurants.

Then Graham heard that there was a bakery nearby that was throwing away tons of mouldy bread every week, and he found out that you can raise maggots on mouldy bread with none of the smells of meat-based production. That was another input to the system, transforming waste into a resource. A further initiative that is not on the diagram is a smokehouse to add value to the fish. This time he worked with the people returning from conflict zones, who often come back with post-traumatic stress disorder (PTSD) and sometimes disabilities (shocking numbers of them actually end up in prison). Whatever we might think about the rights and wrongs of the Iraq War, these are people who have made a big sacrifice and deserve much better treatment. So he was involving them in this system, which was restorative and rehabilitative. What is wonderful about this is that he did not plan it all at the beginning – he let it evolve over time and he used ingenuity. Every time he saw something leaking out of the system or there was something that he needed to buy in, he would think what he could add to that system to create more value, transforming waste into resource. Just as real biological systems increase in diversity and resilience over time, there is a real sense that the more this grows the more the number of possibilities increases. He has been resourceful with the idea of waste in every sense, including what is arguably the most deplorable form of waste and that is underutilised human resources. He managed to reintegrate three marginalised groups (people with disabilities, reforming heroin addicts, and people from conflict zones) in a scheme that is regenerative.

I want to compare some of the characteristics of conventional human-made systems with biological systems. The Cardboard to Caviar Project demonstrates the key characteristics of biological systems: ours tend to be simple and disconnected, whereas biological systems are complex and interconnected. Ours tend to be linear and wasteful; biology is closed-loop and zero waste. Ours tend to be resistant to change; biology tends to be adaptable to constant change. Biology does not use long-term toxins, it runs on solar income, it is optimised as a whole system, and, really importantly, it is regenerative rather than extractive. The characteristics of ecosystems are a good guide to the key transformations that we need to bring about if we are going to shift from the industrial age to what I would call the 'ecological age' of humankind.

The Mobius Project

We have been developing ideas around a productive greenhouse with a restaurant inside for an inner-city area. There is a very successful example in the Netherlands called 'De Kas', where you eat in amongst the plants, and it may be as little as a few hours between the crops being on the plants and their being on your plate – incredibly fresh. We thought it would be great if we could have an anaerobic digester in this building, so we would collect biological waste from the local urban area and then compost it to create methane to create energy and fertiliser from the sludge. We wanted to also treat waste water using a technology called 'living machines', which uses micro-organisms and plants to treat waste water. After the fantastic Cardboard to Caviar Project we were keen to have a fish farm that would be fed from the vegetable waste from the restaurant and would in turn supply fish to the kitchen. We also heard that spent coffee grains are actually an ideal substrate for growing mushrooms (there is still a huge amount of value in spent coffee grains) and so we thought this was another addition. Just for fun we chose a really challenging site – Old Street roundabout – it is a complete eyesore and it is surrounded by four lanes of traffic so no one can get to the middle (Figure 6.7(a)). Since the implementation of the congestion charge nearly all the traffic is from east to north, so with some modest changes we could transform that urban space (Figure 6.7(b)). The idea here is to bring together cycles of food, energy, water, and waste in a way that allows the output from one to become the input for another part of the system. It is moving to a zero-waste, highly productive system that also creates a fantastic amenity. This is a way to rethink the whole metabolism of cities and look at these vast linear flows of resources and transform them from problems into massive solutions.

(a)

(b)

FIGURE 6.7 The Mobius Project. Old Street roundabout in London, (a) without and (b) with implementation of a productive greenhouse scheme. (© Exploration Architecture.)

The Sahara Forest Project

Finally, I would like to discuss the Sahara Forest Project. This was an opportunity to think about integrated solutions. There has been a tendency for designers and environmentalists sometimes to address one problem at a time, whether that is desertification, or climate change, or water shortages. We had a hunch that there would be a way to address multiple challenges simultaneously if we were to think in a really integrated way.

We were quite surprised to learn that a lot of the world's deserts were vegetated a fairly short time ago. When Julius Caesar arrived in North Africa, he found a wooded landscape of cedar and cypress trees. To give you some idea of the abundance that existed, Pliny wrote about the abundance of animals and fruits in the forests, and recorded that for the opening party of the Colosseum in Rome, some 9,000 elephants, panthers, and bears were transported from North Africa to the Colosseum to fight Christians in gladiatorial combat. Say what you like about the Romans, but they knew how to throw a good party. What the Romans started in North Africa was a very extractive model of land use. They cleared the forests to create intensive farming, and for about 150 years North Africa provided vast quantities of grain for the Roman empire, but after that point the Romans found that they had substantially degraded the landscape. It had become salinised; it had lost most of its nutrients that had effectively been washed out into the ocean through the collective digestive system of the Roman empire. They had even changed the rainfall patterns. North Africa used to have quite a plentiful rainfall regime and, by the time they had finished, it was really a desert. That process has continued until the present day, and generally we have continued to adopt this very extractive model of land use.

The amount of energy that falls on the world's deserts is phenomenal and if we could capture even a small amount of that, it would go a long way towards addressing our energy needs. It is reasonably easy to work out the amount of energy we receive as a planet from the Sun every year. If you compare that with the amount of energy we use, we receive roughly 6,000–7,000 times as much energy as we use. I am not suggesting that this shift from the fossil fuel age to the solar age is going to be an easy one – it is not – it is going to be very challenging. But that statistic of 6,000 to 1 does at least suggest that it is possible. It is a challenge to our ingenuity, but it is possible.

Photosynthetic activity around the world has changed a lot over time. What is interesting is that the boundaries of deserts move back and forth over the course of each year. That raised the question for us as to whether we could intervene at those boundary conditions and act to halt or even reverse desertification. One of the reasons why I think this is important is because most biologists agree that if you look at the evolution of life on planet Earth it was really the colonisation of the Earth by plants that helped create the benign climate we currently enjoy. The converse is also true; the more plant matter we lose, the more the deserts grow and the more that is in turn likely to exacerbate climate change and lead to further desertification.

A fairly alarming study was carried out by the Complex Systems Institute in the USA, looking at the link between the food price index and occurrences of civil unrest. They mapped all the significant incidents of civil unrest around the world over the course of about 10 years. There are two main clusters and very close links between the spikes in the food price index and in the occurrences of civil unrest. People sometimes feel uncomfortable when you talk about examples of civil unrest or even violence and you try to find the reasons that lie behind it – as if you are trying to excuse criminality. This is not about excusing civil unrest. It is about understanding the chain of causality that will make civil unrest more likely to occur. It is pretty clear that, when the food price rises above a certain level, civil unrest is almost inevitable. Also the oil price tends to spike about three months before the food price, and that is because most of our food production is incredibly dependent on fossil fuels. So there is an urgent need to break the link between fossil fuels and food production. This is what we have been trying to do on the Sahara Forest Project: developing a way to produce agriculture at large scale while revegetating areas of desert and just using solar energy.

If you are interested in biomimicry and you are working in a desert location, then there is a lot you can learn from the organisms that have already adapted to life there. The Namibian fog-basking beetle is a fantastic example (Figure 6.8(a)). It has evolved a way of harvesting its own fresh water in a desert. The way it does this is that it comes out of its

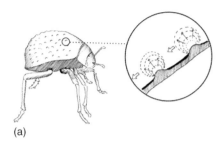

(a)

(b)

FIGURE 6.8 The Sahara Forest Project. (a) A sketch of the Namibian fog-basking beetle harvesting its own fresh water in a desert environment. (© Exploration Architecture.) (b) CGI of the entire scheme, with hedges of greenhouses and concentrated solar power plants. (© Sahara Forest Project.)

hiding place at night, climbs to the top of a sand dune and, because it is matte black, it is able to radiate heat out to the night sky and become slightly cooler than its surroundings. When the moist breeze blows in off the sea you get these droplets of water forming on the beetle's back. Then just before sunrise it tips its shell up, the water runs down to its mouth, it has a good drink, and it goes off and hides for the rest of the day. Not a great quality of life, but it is a clever trick. The ingenuity, if you could call it that, goes even further because, if you look closer at the beetle's shell, it has these bumps on it which are hydrophilic – they attract water – and between them is a waxy finish which repels water. The effect of this is that as the droplets form on the bumps they stay in tight spherical form, which means that they are much more mobile than they would be if there were just a film of water over the whole of the beetle's shell. So, even when there is only a small amount of moisture in the air, it is still able to harvest it effectively and funnel it down into its mouth. This is a great example of what biomimicry can offer in a resource-constrained environment. That is why I think biomimicry is such a powerful source of ideas and very relevant to the kind of challenges we are going to be facing over the next few decades where we're going to come up against resource constraints. It is reassuring to know that we can learn from countless organisms that have already adapted to those conditions.

The Sahara Forest Project involves three core technologies. One is a greenhouse that is cooled and humidified with seawater inspired by the beetle, the second is forms of solar energy, and the third is desert revegetation techniques. Let's say the temperature of the desert at night is $30\,^{\circ}$C. The temperature of outer space is $-273\,^{\circ}$C. With a clear sky you can get a black surface to radiate heat out to outer space. The ancient Persians knew this, and this is how they made ice in the desert thousands of years ago. They would put down a layer of straw as insulation, and then place shallow ceramic trays with a black glaze on top and pour in a thin layer of water. That black surface radiating to outer space was enough to make the water freeze. We are proposing something similar for our greenhouse. We have a double-layer roof and we bring seawater into evaporators, which produce cool, humid growing conditions for the crops. At night we drop the lower layer to create a large roof void. We still evaporate water into this space, and the roof radiates heat to the night sky and forms a condensation surface much like the beetle's shell. Freshwater condensation forms on the underside, runs down to a tank, and is stored for use during the day.

When we think about biology we often think of it as being all about competition. There is certainly a lot of competition in ecosystems, but it is now thought to be less significant than it once was. If you look at mature ecosystems, you are very likely to find examples of amazing symbiotic relationships, where organisms have hooked up for mutual benefit. That is an important principle of biomimicry, looking for ways to bring technologies together in synergistic conditions. We looked around for technologies that would work well with the greenhouse and we settled on concentrated solar power (CSP). This uses mirrors to focus the Sun's heat to create electricity. CSP and the greenhouse have some very interesting synergies. Firstly they both work very well in hot sunny deserts. Secondly CSP can be about 10% more productive if it has sea-water cooling, and we can make use of most of that waste heat to evaporate seawater and create more freshwater. Possibly the most interesting synergy is that the shade created by the mirrors makes it possible to grow a whole range of crops and plants that would not grow in direct sunlight in deserts because the heat is too intense. The idea with this project was to create a long hedge of greenhouses with concentrated solar power plants at intervals along the way. Figure 6.8(b) shows how the scheme would look at large scale. One can see the sea in the top right, the hedges of greenhouses with crops inside, the concentrated solar power at intervals along the way and then, on the left, the image indicates pioneer species being used to push back the boundaries of the deserts. This is a model for how we could create zero-carbon food in some of the most water-stressed parts of the planet whilst also producing abundant renewable energy and locking up carbon back in desert soils.

That is where we were in 2010, and we managed to get a lot of positive publicity for this. We were hoping we would get a phone call from a philanthropic tycoon asking for a project, but it did not quite work out that way. So I started working through my address book and sent this to quite a few people, including Frederic Hauge of a Norwegian NGO called The Bellona Foundation. The foundation became a partner in the project, and they brought with them amazing funding and political connections. They helped us put the project in front of the right people, which led to a deal for three feasibility studies in Jordan and, in a separate deal, a further feasibility study in Qatar. During these studies we looked at a number of other technologies that could work well in our synergistic cluster, including algae for biofuels and halophytes (which are crops that can be grown directly in seawater and produce food and fodder). We looked at evaporator hedges, which would take the brine from the greenhouse and evaporate it out towards a saturation point so that we could then produce dry salts and extract useful compounds and elements from the brine to use as fertilisers. We have connected all these technologies, trying to achieve what I referred to earlier as a complex interconnected system that moves towards being highly productive and zero-waste. In principle we are using what we have a lot of (sunlight, seawater, and carbon dioxide) to produce more of what we need (biomass, oxygen, electricity, crops, and materials).

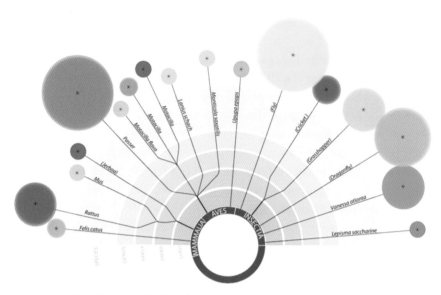

FIGURE 6.9 Biodiversity on the Sahara Forest Project. (© Exploration Architecture.)

We built a version of this in Qatar that was opened during the climate change talks in Doha in 2012. In addition to all the technical aspects of the project, one of the things we monitored was the biodiversity. We wanted to demonstrate that this could have a powerful restorative effect. We looked at the animal species that appeared, the main ones being mammals, birds, and insects (Figure 6.9). The first animals to appear were flies, as nearly always seems to be the case, and then as soon as the first plants arrived on site we had house sparrows, and pretty soon after that we had grasshoppers and crickets, and then there were wagtails. The number of insects was increasing, and we had the first butterfly appear. Then we had the first problematic species, rats, and this was because we were being a bit careless with the seeds that we were leaving around. Then we found mice and more birds appearing, and the number of insects was increasing all the time. Then, literally three days after the algae ponds had been filled, the first dragonfly appeared, and bear in mind that this was in the middle of the desert, a long way from the nearest bit of vegetation. Then a feral cat appeared and the number of rats started going down. When the halophytes were established we started getting more birds, including a relatively rare hoopoe, and then we had Rufous-tailed shrikes, more insects, more types of wagtail, and so on. And finally we had a little indigenous mammal, a jerboa, which is like a mini-kangaroo. This was all achieved in eight months, on a site that is 100 metres by 100 metres. On a bigger scale, over a longer period of time, this regenerative effect would be even more pronounced.

To conclude now, we are next working in Jordan. We often think about Jordan as a haven of peace surrounded by troublesome neighbours, but actually it is in a very precarious situation. Jordan imports 95% of its energy, it is one of the most water-poor countries in the world, and it has a growing population – swelled with a lot of people fleeing conflicts, so it could very easily unravel into conflict. Projects such as ours could really deliver a lot in terms of improving their food, water, and energy security.

Conclusions

I mentioned some of the frustrations of those working in the environmental field, the frustrations with economists and the deniers, or with the doomsters, but I do not think we should get too distracted by the doomsters or the deniers, or even by the optimists or the pessimists. We do need scientists to keep us informed, and in some cases to create justifiable fear. But, as designers, I think we have a different role. Rather than creating fear about the future, the future is something we should fearlessly create. I do think we need a more sophisticated notion of economics. We need to bring the *nomos* of economics and the *logos* of ecology together – management based on knowledge – creating solutions that deliver optimum long-term value rather than maximum short-term return.

The biomimicry solutions I have described are just a tiny fraction of what exists, in a sourcebook that we have only just begun to understand: crabshells and gridshells, high-strength polymers, high-strength composites, super-efficient structures – 3.8 billion years of research and development, 3.8 billion years of brilliant solutions – illuminated by previously unparalleled scientific knowledge, and facilitated by previously unimaginable digital design tools. Designers have never had such an opportunity to rethink and design solutions fit for the next billion years.

Returning to those three big challenges laid out at the beginning – radical increases in resource efficiency, going from linear to closed loops, and the passage from a fossil fuel to a solar economy, I firmly believe those are possible and that this future we are creating, the ecological age, could be a really wonderful abundant future. Of course, some cynics might scoff at the feasibility of shifting from fossil fuels to a solar economy, and scoff at the idea of zero waste. Importantly, we know that this is not the stuff of fantasy, because the natural word is living proof of the possibility. And it is not just an ordinary possibility, but a wonderfully rich, abundant, and regenerative possibility.

Reference

Eisner, T., & Aneshansley, D. J. (1999). Spray aiming in the bombardier beetle: Photographic evidence. *Proceedings of the National Academy of Sciences of the United States of America*, 96, 9705–9709. doi: 10.1073/pnas.96.17.97051.

7 Economic Development

HA-JOON CHANG

Why Are Poor Countries Poor?

George W. Bush once famously said that the problem with the French is that they do not have a word for 'entrepreneurship'. Mr Bush's French may not have been up to scratch, but he was articulating a fairly common Anglo-American prejudice against France as an un-dynamic and laid-back country – full of meddling bureaucrats, pompous waiters, and sheep-burning farmers. This conception of France turns out to be wrong, as will be shown later, but the perspective behind his statement is widely accepted – you need entrepreneurial people to have a successful economy.

In this view, the poverty of the developing countries is attributed to the lack of entrepreneurship. So, when people from rich countries go to a developing country and come across people idling away their days in cafes, they say 'Look at all those men sitting around having their eleventh cup of mint tea of the day and smoking their hookahs; these countries really need more movers and shakers in order to pull themselves out of poverty.'

However, anyone who is from, or has lived for a period in, a developing country will know that developing countries are teeming with entrepreneurs – men, women, and children trying to sell everything you can think of, and things that you did not know could be sold and bought. In many poor countries, you can buy a place in the queue for official businesses (sold by professional queuers), a couple of young men to be extra passengers in your car so that you can drive in the car-pool lane, the right to set up a market stall on a particular street corner (perhaps sold by the corrupt local police chief), or even a patch of land to beg from (sold to you by the local thugs).

In contrast, most citizens of rich countries have not even come near to becoming an entrepreneur. They mostly work for a company, some of them employing tens of thousands, or even millions, of people, doing highly specialised jobs, and in the process realising someone else's entrepreneurial vision.

According to a study by the OECD, the club of rich countries (OECD, 2009), 30–50% of the non-agricultural workforce in most developing countries is self-employed (the ratio tends to be even higher in agriculture). In some of the poorest countries, the ratio of people working as one-person entrepreneurs can be way above that: 66.9% in Ghana, 75.4% in Bangladesh, and a staggering 88.7% in Benin. In contrast, only 12.8% of the non-agricultural workforce in developed countries is self-employed. In some countries, the ratio does not even reach one in 10: 6.7% in Norway, 7.5% in the USA, and 8.6% in France – so it turns out that Mr Bush's complaint about the French was a classic case of the pot calling the kettle black.

So, even excluding the farmers, which would make the ratio even higher, the chance of an average developing-country person being an entrepreneur is more than twice that for a developed-country person (30% vs. 12.8%). The difference is a factor of 10, if we compare Bangladesh (including farmers) with the USA (75.4% vs. 7.5%). And in the most extreme case, the chance of someone from Benin being an entrepreneur is a whopping 13 times higher than the equivalent chance for a Norwegian (88.7% vs. 6.7%).

Moreover, even those people who are running businesses in the rich countries need not be as entrepreneurial as their counterparts in the poor countries. For developing-country entrepreneurs, things go wrong all the time. There are power cuts that screw up the production schedule. Customs do not clear the spare parts needed to fix a machine, which has been delayed anyway due to problems with the permit to buy US dollars. Inputs are not delivered at the right time, as the delivery truck broke down – yet again – due to potholes on the road. And the petty local officials are bending, and even inventing, rules all the time in order to extract bribes. Coping with all these obstacles requires agile thinking and the ability to improvise. An average American businessperson would not last a week in the face of these problems, if they were made to manage a small company in Maputo or Phnom Penh.

So we are faced with an apparent puzzle. Compared with the rich countries, we have far more people in developing countries (in proportional terms) engaged in entrepreneurial activities. On top of that, their entrepreneurial skills are much more frequently and severely tested than those of their counterparts in the rich countries. Then how is it that these more entrepreneurial countries are the poorer ones?

The most popular answer to this in the last decade or so has been that the poor people in developing countries lack some critical inputs, especially credits, with which they can exercise their entrepreneurship. This thinking is what is behind the celebrated microfinance movement – lending small sums of money to poor people so that they can use it to start businesses and stand on their own feet. Or at least that is the theory.

Despite the high hopes it generated, the spread of microfinance has not done much to lift people out of poverty. There are many reasons behind this, including the excessively high rates of interest charged – typically microfinance institutions (MFIs) charge 40–50% annual interest rates, while some of them charge 100% or even 200%. However, one key reason is the low productive capabilities of the individuals taking on the loans.

Let me explain this with an example. In 1997, the Grameen Bank of Bangladesh teamed up with Telenor, the Norwegian phone company, and gave out microloans to women to buy a mobile phone and rent it out. Initially, these 'telephone ladies' made handsome incomes – $750–1,200 per year in a country whose annual average per capita income was around $300. However, by 2005, the business got so crowded that their average income was only around $70 per year, even though the national average income had gone up to over $450 in the meantime.

Of course, this problem would not have existed if new business lines could be constantly developed. So, for example, if phone renting becomes less profitable, you could maintain your level of income by manufacturing mobile phones or writing the software for mobile phone games. You will have, of course, noticed the absurdity of these suggestions – the telephone ladies of Bangladesh simply do not have the capabilities to move into phone manufacturing or software design. This example shows that microfinance has failed to work because, given their low productive capabilities, there is only a limited range of businesses that the poor individuals in developing countries can take on, which quickly gets crowded.

The Current State of Development Discourse:
The Neglect of Production

The short story about microfinance that we described in the previous section is very symptomatic of the current state of development discourse.

Until the 1970s, there was a general consensus that economic development is essentially about the transformation of the abilities to produce – that is, productive capabilities – although people fiercely debated how to achieve it. In fact, most of us still hold such a view of development at the instinctive level. For example, in refusing to describe oil-rich high-income countries as 'economically developed', we are implicitly saying that achieving high income through a resource bonanza is not 'economic development'. At the other extreme, following the Second World War, the German income level fell to that of Peru or Mexico, but few people at that time argued that Germany should have been re-classified as a 'developing' country, because people knew that Germany still had the necessary productive capabilities to regain its pre-war level of living standards quickly, which it did in about 10 years. These examples show that we are implicitly saying that economic development is about the capability to produce, rather than about simple command over resources.

However, during the last three decades, the dominant view has become that economic development is basically about poverty reduction and a more widespread provision of basic needs, as best illustrated by the United Nations' Millennium Development Goals (MDGs), which include nothing about production.[1]

Behind this radical transformation in our discourse on economic development have been three intellectual trends – the rise of neo-liberalism, the humanist reaction to the earlier production-oriented development discourse, and the emergence of the discourse of post-industrial service economy.

---

[1] The MDGs are (1) eradicate extreme poverty and hunger; (2) achieve universal primary education; (3) promote gender equality and empower women; (4) reduce child mortality; (5) improve maternal health; (6) combat HIV/AIDS, malaria, and other diseases; (7) ensure environmental sustainability; and (8) develop a global partnership for development.

The most important of these trends is the rise of neo-liberalism – an economic doctrine that argues that the best economic policy is a slightly updated version of the nineteenth-century liberal policy of laissez-faire. There are a number of reasons why this has led to the neglect of production. First, neo-liberalism is based on, although not synonymous with, neoclassical economics, which is basically about market exchange and has little to say about production; the 1992 winner of the Nobel Prize in economics, Coase, criticised it for being about 'lone individuals exchanging nuts and berries on the edge of a forest' (Coase, 1992, p. 718). Second, in neo-liberal – and neoclassical – economic theory, the issues surrounding the development of productive capabilities are assumed away, as it takes these capabilities as given. For example, in neoclassical trade theory, all countries are assumed to have the same capabilities to use any technology – so, in this theory, if Guatemala is not producing things like BMWs, it is not because it cannot, but because it should not (and does not want to), given that the technology for their production uses a lot of capital and very little labour, when Guatemala has a lot of labour and very little capital (of course, in proportional terms). Third, believing in the power of the free market, the neo-liberal discourse criticises any attempt to deliberately enhance productive capabilities through public policy intervention – which was at the heart of the early development thinking – as being at best futile and at worst counter-productive.

Another intellectual trend behind the decline of the interest in production has come from the left of the political spectrum. Since the 1980s, there has been a 'humanist' reaction – represented by Amartya Sen's Capability Approach – against what was seen as the collectivist, materialistic biases of the early development discourse. The early development theories were mainly about transforming the productive structure through policies affecting aggregate variables, like investments and total labour supply. In applying these theories, individuals were forgotten and, worse, repressed in the name of the greater good, called economic development. This has made the 'humanists' emphasise the need to enhance individual capabilities through health, education, and empowerment, as reflected in the MDGs.

Last but not least, the rise of the discourse on the post-industrial service economy has also helped indirectly to weaken our interest in production. According to this view, rising income has brought about the shift of demand towards services, making material production increasingly less important. The advocates of this position cite cases like Switzerland and Singapore as examples of service-based prosperity. The issue of productive development is almost completely ignored in this discourse, as it is implicitly assumed that even the poorest countries possess the capability to compete in services – any country can run call centres, the implicit assumption goes, even though it may not be able to run a car factory. The recent Indian success story with service export is often presented as the proof of this.

Now, let me critically examine this trinity of arguments that has led to the neglect of production in development discourse today – the neo-liberal view, the humanist view, and the post-industrial discourse.

Regarding the neo-liberal view, there is ample historical evidence showing not only that industrialisation is necessary for economic development but also that it does not happen automatically through market forces.

First of all, neo-liberal policies have failed to generate economic development in the last three decades. In the last three decades, Sub-Saharan Africa (SSA) and Latin America have applied neo-liberal policies most diligently, partly voluntarily but mostly under compulsion from the World Bank and the International Monetary Fund, on which they depended financially. The per capita income growth rate fell from 3.1% in Latin America and 1.6% in SSA during the 'bad old days' of state interventionism in the 1960s and 1970s to 0.8% and 0.2% in the next 30 years of neo-liberalism (1980–2010). There have been growth pick-ups in both regions in the last decade or so, but this was largely due to a commodity boom, rather than the developments in productive capabilities.

Moreover, almost all of today's rich countries – including Britain and the USA that are known as the countries which invented free trade and the free market – developed by using state intervention, like trade protectionism, industrial subsidies, state-owned enterprises, and regulations of foreign direct investments, as I have shown in my work (Chang, 2002, 2007).

As for the humanists, their problem is not that they are not interested in raising productive capabilities – they are. Their problem is that they

try to do it mainly by making individuals more capable through investment in health and education. However, there are only so many productive capabilities that can be developed through improvements at the
individual level.

First of all, developments in productive capabilities in the modern
world mainly occur at the level of business enterprises, rather than at
the individual level. In other words, what really distinguishes the USA or
Germany, on the one hand, from the Philippines or Nigeria, on the other
hand, are their Boeings and Volkswagens, and not their economists or
medical doctors (which the latter countries have in quite large quantities).

Second, in addition to the development of business enterprises, you
need to develop a series of collective institutions that encourage and help
different economic actors work together – capital–labour collaboration
within firms, cooperation among firms within and across sectors,
government–business interaction (including, but not just, industrial
policy), industry–academia partnership, and so on.

A simple way to see the importance of these collective dimensions of
productive capabilities is to recall how, when an engineer from a poor
country emigrates to an economically developed country, their productivity increases enormously. The neglect of these collective dimensions of
productive development is an important gap in the 'humanist' tradition.

The post-industrial economy discourse also has some critical weaknesses.

First of all, the increasing share of services in output does not mean
that material production is not important any more. We are in fact
consuming an ever-increasing amount of material goods, and many
countries are producing even more material goods than ever in absolute
terms. The rise in the share of services happens mainly because services
are becoming relatively more expensive, given the faster rise of productivity in the manufacturing sector (see Chang, 2010, Ch. 9; 2014, Ch. 7).

Second, a lot of recent productivity growth in service sectors like
finance and retail has been illusory. The high productivity growth of
the financial sector has been based on dubious asset valuation, market
rigging, and sometimes even fraud, and has come at huge costs to the rest
of the economy, as seen in the 2008 global financial crisis. A lot of the
increases in retail service productivity in countries like the USA and
Britain have been bought by lowering the quality of the retail service

itself – fewer sales assistants, longer drives to the supermarket, lengthier waits for deliveries, and so on.

Third, most high-value services – finance, engineering, IT services, consulting, etc. – mainly sell to the manufacturing sector, so they cannot prosper without a strong manufacturing base.

As for the supposed service-based success stories, it must be first pointed out that Switzerland and Singapore – the supposed models of service-based prosperity – are in fact literally the two most industrialised countries in the world. According to the statistics from the United Nations Industrial Development Organization (UNIDO), at $10,110 per person, Switzerland produced the largest amount of manufacturing output in the world, measured by manufacturing value-added (MVA), producing 9.5 times more manufacturing output than China, the supposed 'workshop of the World', which had per capita MVA of $1,063. Singapore, at $8,966, ranked no. 2 in the world, followed by Finland ($8,097, no. 3), Sweden ($7,419, no. 4), and Japan ($7,374, no. 5).

India's service-based success story is also highly exaggerated. Between 2004 (until then India had a deficit in service trade) and 2011, India recorded a service trade surplus equivalent to 0.9% of GDP, which covered only 17% of its merchandise trade deficit (5.1% of GDP). This means that, unless it increases its service trade surplus by a factor of 6 – an implausible scenario, given that its service trade surplus has *not* even been on a firm rising trend – India cannot maintain its current pace of economic development without a serious balance-of-payments problem.

Does It Matter That We Neglect Production?

Now, we may ask whether it really matters that we neglect production – or more precisely the issue of productive capabilities development – in defining economic development. My answer is that the neglect really matters, as it has resulted in a number of negative outcomes.

First, the neglect of productive capabilities has made a lot of people think that what countries produce to earn their incomes does not matter – 'It does not matter whether you produce potato chips or micro-chips,' to borrow a famous expression from the industrial policy debate in the USA in the 1980s. This has made a lot of developing countries complacent about their dependence on primary commodities or cheap assembly. However, in the long run, different economic activities give different scope for output expansion, productivity growth, and technological progress, so, even from a purely growth-oriented point of view, this line of thinking is highly problematic.

Second, the neglect of productive capabilities has also meant that our assessment of economic policies has acquired short-term biases. Policies that reduce current consumption with a view to increasing long-term productive capabilities, such as infant industry protection or policies to deliberately increase investments, are these days too easily dismissed.

Third, the neglect of the collective dimension of productive development has made people ignore the issue of how to develop modern firms and other institutions that are central to productive development. There is of course a lot of talk of 'private sector development', but it is mainly a 'negative' agenda in the sense that reducing the involvement of the public sector, through deregulation and cutting taxes, is seen as automatically leading to enterprise development.

Fourth, the neglect of production has also led to a very partial view of our individual well-being. Envisaging people as consumers rather than producers (workers), issues of employment, the quality of jobs, and workplace welfare have disappeared from our public policy agenda.

The Way Forward

Given these negative consequences of neglecting production, we need to re-construct our discourse on economic development by bringing production back in – especially the issue of the development of collective productive capabilities. This is not to suggest that we simply go back to the older collectivist, 'productionist' tradition in development thinking. We need to incorporate the more recent theoretical developments and real-world concerns.

First, we need to incorporate the insights from the humanist tradition and pay greater attention to non-material aspects of development – like freedom, equality, solidarity, and community.

Second, emphasising production should not mean neglecting individuals, as was often the case in the older productionist tradition in development thinking. We need to take individuals more seriously. However, we need to see them not just as consumers (as in neo-liberal economics), nor just as citizens with entitlements (as in the humanist approach), but also as producers, including the important dimensions of life employment and the contents of their work.

Third, reflecting the recent developments in the economics of technology and institutional economics, our understanding of production itself needs to be improved, with attention paid not just to technologies, but also to the role of organisations, institutions, culture, and even politics in the process of the development of productive capabilities.

Last but not least, environmental sustainability has to be incorporated into our thinking. Contrary to conventional wisdom, development of productive capabilities, especially in the manufacturing sector, is crucial in preventing and adapting to climate change. Given that they are the ones with the necessary technological capabilities, the rich countries need to further develop their productive capabilities in the area of green technologies. Even just to cope with the adverse consequences of climate change, developing countries need to further develop technological and organisational capabilities, many of which can be acquired only through industrialisation (see Chang, 2014, Ch. 7).

Developing and enriching the traditional 'productionist' view of economic development with the addition of these new elements will not be easy, but it is absolutely necessary, if we are to promote more rapid but economically and environmentally sustainable economic development.

References

Chang, H.-J. (2002). *Kicking Away the Ladder – Economic Development in Historical Perspective.* London: Anthem Press.
  (2007). *Bad Samaritans.* London: Random House.
  (2010). *23 Things They Don't Tell You about Capitalism.* London: Penguin.
  (2014). *Economics: The User's Guide.* London: Penguin.
Coase, R. H. (1992). The institutional structure of production. *American Economic Review,* 82(4), 713–719.
OECD (2009). *Is Informal Normal? Towards More and Better Jobs in Developing Countries.* Paris: Organisation for Economic Co-operation and Development (OECD).

8 Technology Development

HERMANN HAUSER

In this essay, I will be addressing the topic of technological development through time. However, because of my background, I will concentrate largely on what many immediately think of as technological development, namely the development of computer technologies that many of us have observed over our lifetimes.

Professor Richard Lipsey, Professor of Economics at Simon Frazer University in Vancouver, studied what he called general-purpose technologies (GPTs) in some detail and defined them as technologies that really affect the entire economy, not just a part of it. This leads to the obsolescence of all technologies and skills, causing temporary unemployment, and it is clear that these transitions from one GPT to another might take a long time, require new infrastructure, and be quite expensive in terms of the learning cost.

From the beginning of time until the first century BC there have been seven:

1 the domestication of plants in the Neolithic Age
2 the domestication of animals
3 the smelting of ore in the eighth millennium BC
4 the wheel
5 writing
6 the Bronze Age
7 the Iron Age

In the next 19 centuries there were another nine:

8 the waterwheel
9 the three-masted sailing ship
10 the printing press
11 the factory system
12 the steam engine

13   the railways
14   the iron ship
15   the internal combustion engine
16   electricity

But then comes the twentieth century, and in just one century we got eight of these general-purpose technologies that change everything:

17   the automobile
18   the aeroplane
19   mass production
20   computers
21   lean production
22   the Internet
23   biotechnology
24   nanotechnology

I would add Artificial General Intelligence as the 25th GPT that may be the defining technology of the twenty-first century. It may possibly be the last GPT, as an Artificial General Intelligence could be applied to all outstanding problems.

So new technologies can change our lives in a very fundamental way. To illustrate this, let me introduce Rudolf Pižl. He was a small factory owner in Vienna, who produced leather belts for the industrial revolution. When I was a small boy I watched him buy cow hides, cut them into strips, glue them into belts and pre-stretch them before they were sent to the factories. He was my grandfather.

Factories in his day were laid out with a steam engine at one end. It produced all the power needed for the entire factory which needed to be distributed to the individual workstations, for example looms. This was done by running axles along the length of the factory just below the ceiling (Figure 8.1). These axles had wheels which were connected by my grandfather's belts to the individual workstation on the factory floor.

But then came electricity, and that caused a substantial change in the way a factory floor was organised. Instead of needing belts to access mechanical energy, electricity was distributed through wires along the factory floor. There was no need for a steam engine because there was a central electricity power station far away providing the power through a distribution network, for many factories, not just one.

FIGURE 8.1 A factory floor powered by a steam engine.

The consequence was that many new jobs were created and new skills required to produce the motors, to lay the electric cables, and more recently to produce factory robots, but these were skills and jobs that were very different from the ones previously required. However, it took a long time for these changes to be fully adopted, so my grandfather and others affected like him could change to a new business, in his case distributing rubber belts for combine harvesters and fan belts in cars. This business went on for another generation.

Now let us focus on GPT number 20, computers, and let me illustrate the differences between the rates of change of the previous technological waves and the rates of change in the computing field. Figure 8.2 shows the super-exponential rise of computer power that can be bought for $1,000.

There are six waves that can be distinguished in the development of computing.

Like every good computer scientist, I'll start my six waves with the zeroth wave: the EDSAC. The EDSAC was developed in Cambridge and was the world's first user computer. All computers beforehand were either developed for the military to do very specific tasks or were developed by computer science departments to prove a particular architecture.

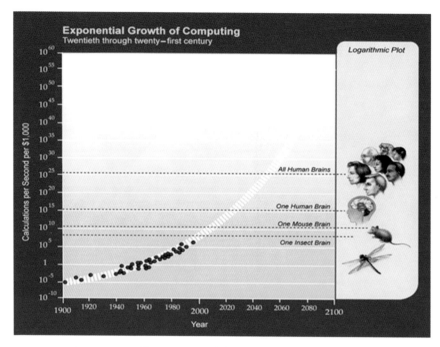

FIGURE 8.2 Super-exponential growth in power of computers costing $1,000 according to Ray Kurzweil. (Courtesy of Ray Kurzweil and Kurzweil Technologies, Inc.)

In Figure 8.3, Maurice Wilkes, who designed the EDSAC, kneels in front of the mercury delay line. EDSAC stands for Electronic Delay Storage Automatic Calculator; not computer, because a computer at that time was a person, not a machine.

The mercury delay line was the main storage device. There was no other way of storing bits at that time, so this ingenious device was designed. It worked like this: when the mercury was banged at one end, a wave was propagated along it and a microphone at the other detected the waves. When there was a wave it was a one; when there was none in a given time interval it was a zero. A single mercury tube could store about 500 bits.

The computing elements were developed using vacuum tubes. The EDSAC managed 650 instructions per second. Today everything is measured in MIPS (million instructions per second), so you may think this was very slow, but it represented an increase in computing power of

FIGURE 8.3 Maurice Wilkes (kneeling) with the EDSAC. (Courtesy of the Computer History Museum.)

1,500×. This was the largest increase in computing power that users have ever had, and is likely to remain the largest increase in computing power from one year to the next that we ever will have.

Professor David Wheeler invented the subroutine on this machine, which has become one of the most fundamental ideas in software architecture.

The EDSAC also contributed to three Nobel Prizes, one for Max Perutz and John Kendrew for the structure of myoglobin, one for Andrew Huxley for understanding how neurones work, and the third for Martin Ryle and Antony Hewish for radio astronomy.

Let's now consider the user interface in each of these waves, as it is the key to the wider adoption of computers in society. The user interface for the EDSAC was very primitive. Users had to punch their programs on to paper tapes and queue up, to put them into the paper tape reader. That is where the term 'job queue' comes from. It was the only way to put a program into the machine. The program had to be in machine code, which is very obscure. The invention of the subroutine made things a little easier, but it was still very hard work. The output was via a printer.

Now we come to the six waves of computing:

1   the mainframe
2   the minicomputer
3   the workstation
4   the personal computer
5   the smartphone and the cloud
6   the Internet of Things and machine learning

Let us start with the first wave: the mainframe. Figure 8.4 shows the iconic product of that mainframe era: the IBM 370. As a PhD student I remember using the IBM 370. Few people used paper tape anymore, as the main modes of data entry were via punched cards or teletype machines. Some of you may remember that whenever you saw a computer on TV in those days they focused in on the tape readers, because they were the only moving part of the computer and as they went around it gave people the feeling that the computer was working.

So what were the characteristics? Well, it was quite an expensive unit at a million dollars. They were sold at the rate of about 10,000 units a year. Because they were so expensive, they were used by many people at the same time. In Cambridge University 200 people would be users of the 370, which required an air-conditioned room to function. There was no

FIGURE 8.4 An IBM 370 mainframe. (Author: Oliver.obi. CC-BY-SA-3.0 license. https://en.wikipedia.org/wiki/File:IBM_System_370-145_und_Bandlaufwerke_2401.png.)

network, and the main use cases were in the headquarters of large organisations, for accounting, and in universities, for scientists.

The dominant company at the end of this wave was IBM. This will be a characteristic of all six waves, that one company becomes dominant. IBM produced the hardware, the software, the application, and the services, and at the end of the life of the machine they would also take it away from you for a fee. IBM was totally vertically integrated. What happened subsequently is sometimes called the horizontalisation of the industry as the vertical stack was split into a number of segments of specialised companies that produced the different components (e.g. silicon chips, processors, boards, systems software, application software, networks, services etc.).

The other players of that era were called the BUNCH: an acronym for Burroughs, UNIVAC, National Cash Register (NCR), Control Data, and Honeywell. But there was also another interesting aspect of that wave which will repeat itself, and that is the lock-in. Often these monopoly players lock users in to their product, which results in high margins for these producers. This is indeed what happened at IBM. Their margins were around 80%. The lock-in was the DASD disk, which only IBM managed to produce. The interface to the disk was proprietary, so none of the competitors could use it. In the end they were forced by the US anti-trust authorities to make it public so that other companies were no longer locked out of that business. Whenever you see an anti-trust investigation of a company, as we'll see later with Microsoft, you can assume that the end of such a wave is near.

So what was the user interface like? Well, at the beginning it was punched cards, then the teletype. But it became much easier to program computers because IBM invented high-level languages such as FOR-TRAN, which stands for Formula Translation for the Scientific Community, and COBOL for business forms for big businesses.

The second wave ushered in the minicomputer. The iconic product for this wave was produced by a company called DEC that grew fantastically from zero to one of the most valuable companies at the time on the back of the PDP-11 product range, and then the VAX 11/780 (Figure 8.5), which became the most popular minicomputer at the time.

What happened, why was this so popular?

FIGURE 8.5 The Digital Equipment VAX 11/780 minicomputer. (Author: Joe Mabel. CC-BY-SA-3.0 license. https://commons.wikimedia.org/wiki/File:LCM_-_DEC_VAX_11-780-5_-_01.jpg.)

Well the price came down by a factor of ten, and that is significant. Not surprisingly the number of units sold went up by a factor of 10 from 10,000 per annum to about 100,000 per annum, and the user base and the use case grew very rapidly. It also meant that smaller companies could now afford computers.

But the dominant company, DEC, is no more, and that is what often happens to a near monopolist towards the end of a wave, because these companies get so good at serving their particular wave that they miss the next wave every time.

The user interface to these minicomputers became more advanced, with a visual display unit and a keyboard. Programs such as Computer-Aided Design (CAD) tools, and high-level programs that were fantastically powerful to help mechanical, electronic, and software engineers to become a lot more productive emerged, and a new hardware component made its entry, the Ethernet. This was the beginning of the Local Area Network (LAN) era, which has lasted to this day.

The third wave brought us the workstation, and again we have a dominant product, the Sun workstation (Figure 8.6). The price came down again by a factor of 10, the number of units went up by a factor of 10, and the use case widened enormously, with this unit being so small that it fits on a desk. Thus it could become a computer for a single user.

The dominant companies were Apollo and Sun, and for a while there was competition, which Sun eventually won, becoming the dominant supplier of workstations. There was innovation on the mass storage front, with small hard disks becoming quite capacious, and the Ethernet could link a few together so people could use not just a single workstation but clusters of workstations.

The user interface was that of the desktop, which was invented at Xerox PARC and later copied by Apple for their desktop user interface, and developed to include a mouse. It was nicknamed the 4M machine as it had

1   1 MIP (million instructions per second)
2   1 megabyte of memory
3   1 megapixel display, which enabled detailed engineering drawings to be displayed on the screen
4   It cost 1 mega-penny, i.e. 10,000 dollars

This brings us to the fourth wave and the personal computer. If you'll allow me, I'll have a nostalgic moment here with the BBC Micro, which was designed and produced by my first company, Acorn Computers. The BBC had taken the far-sighted decision to educate the nation about

FIGURE 8.6 A Sun workstation.

computers, and chose the Acorn machine in competition with six other British firms to be the practical teaching tool of the initiative. It was introduced as part of the BBC computer programme. It is very difficult to imagine that at that time, which was the early 1980s, a television programme at six-o'clock in the evening could become a national phenomenon and that people would rush home from the pub to see it. On the basis of this programme, Acorn's BBC computer became the standard in British schools, and what I am most proud of, even more so than the most successful company that is associated with me, which is ARM, the Advanced RISC Machine company, is that the BBC Micro created a whole generation of software programmers.

We now have a worthy successor to the BBC Micro, the Raspberry Pi, which was developed in Cambridge; a stripped-down, but fully functional, educational computer about the size of a cigarette packet starting at $20, which has now sold over five million units.

But it must be admitted that the iconic product of the PC era was really the ugly-looking IBM PC in Figure 8.7.

What happened? Why did the personal computer become so successful? Well, it is the old story: a tenth of the price (they now sold for $1,000), and the numbers sold skyrocketed, not just one million or 10 million per annum, but now in the hundreds of millions per annum. The use case is a personal use case based on productivity tools and more recently on browsing. We are coming to the end of this wave where the dominant company (Wintel) is actually a duopoly between Intel and Microsoft, with Intel having an 80% market share on the microprocessor side and Microsoft a 90% market share on the operating system side.

FIGURE 8.7 An IBM Personal Computer (PC). (Author: Boffy b. CC-BY-SA-3.0 license. https://commons.wikimedia.org/wiki/File:IBM_PC_5150.jpg.)

The user interface is a scaled-down version of the workstation user interface with a desktop. The main applications are Word, Excel, and PowerPoint, with many other applications as well, but interestingly one application, more than anything else, in the end became the application that every one of you uses most of the time: the Internet browser.

So it is the connectivity with the Internet which provides the main usage of PCs. As an aside, there are extensions to the user interface (UI) like Kinect from Microsoft, a camera that watches what you are doing with a 3D capability. Another UI extension is eye-tracking, which is led by a Swedish company called Tobii. This is quite important because if a computer knows where you are looking, if it tracks your gaze, it can be very helpful in bringing relevant items to your attention.

On the power chart we have moved up another notch. Computers are getting close to the size of the brain of an insect and we are now onto the fifth wave, the smartphone and the cloud.

Steve Jobs, an absolute genius, defined the smartphone with the iPhone. When, at the iPhone launch, he announced to the world 'We've reinvented the phone,' I was sceptical, but he did: he absolutely reinvented the phone.

At that time Nokia had a 40% market share, and I remember talking to some of the Nokia executives after the launch of the iPhone and they all sighed with relief because, they said, 'It is a poor phone.' What they looked at was the radio capability; they completely missed the main point, which was the user interface. It was just not in the mind-set of Nokia to think about a touch interface the way Apple did; that was the stroke of genius.

Why did these smartphones take off, and why are they now the most popular form of computer? More people now use the Internet via a mobile phone than via a PC. Well, it is the same old story: the price has fallen to zero if you are willing to take out a contract. The units are now produced in the billions per annum, not just hundreds of millions, and you can use a phone everywhere you go. It has grown from a phone to be the most popular computing device.

Interestingly, smartphones are based on a new processor, the ARM processor, and we also have new operating systems, which are not from Microsoft. This development started with the Symbian OS, produced by a UK company called Psion, which was adopted by Nokia, but soon Apple and Android became completely dominant.

There is a very interesting company that relatively few people have heard of, called Xiaomi. At the moment the market leader is still Samsung, followed by Apple, but Xiaomi has become the number one brand in China and it is the fastest-growing smartphone company in the world. They sell smartphones for $35, so it is not surprising that they sell well.

How can they do this? They sell them at cost. Why would they sell their mobile phones at cost? They have a different business model: the phone is the means by which they sell other electronics products, apps, and services. This is one of the secrets that we'll come across again with ARM, as sometimes it is the business model, rather than the product, which leads to success.

Now for tablets: do not think that this is a new type of machine. This is actually a mobile phone on steroids and not a shrunk PC. It has the same architecture, the same ARM processor, and the same OS as the phones, albeit with a larger display.

The part of the fifth wave that is absolutely key in terms of the usefulness of a mobile phone is the data centre that sits behind, providing all the information that you want through the Internet.

The main phone usage is voice conversations with other people, but you can also use the mobile phone as a voice interface to the Internet.

The user interface has become quite voice-centric, and we'll see much more development of voice with companies like VocalIQ. This is a Cambridge company, a spin-out from the Engineering Lab that adds dialogue management, which is the final layer to voice recognition, because through dialogue any ambiguities in a question or instruction can be resolved.

The user interface also depends a lot on the amazing explosion of apps. Both Android and iOS have over a million apps now, meeting all kinds of needs.

We now come to the sixth and last wave of computing, which is the Internet of Things and machine learning, and as you'll see in a moment it is actually machine learning that is the more important part here.

Surprisingly we get yet another factor of 10 in the number of units being produced per year. I would have thought once you get to sales of a billion units a year that would be about the limit, as there are only seven billion people on Earth. But a factor-of-10 increase is what is happening.

The Internet of Things will provide a veritable tsunami of data, so the big-data aspect of the Internet of Things is the key to its future success.

I have made the point that the price also went down by a factor of 10, and we arrived at zero dollars in the last wave, so how can you do any better than that? Well, the answer is price actually does not come into it anymore because the Internet of Things part, the computing part of it, is really part of something that you would buy anyway: it is part of a heating system, a part of a car, or part of other appliances. It is an additional function added to something that you need anyway.

The most important aspect of the Internet of Things that I want to leave with you is not that there are so many individual devices, and we are talking about tens of billions soon, but the fact that they are all connected. The brain has 100 billion neurones, but 100 trillion connections. The connections of the Internet of Things will far exceed that. It will be trillions and quadrillions once they are all connected.

Examples of the Internet of Things are the Google car, Google Glasses, and the health sensors that are appearing at some speed, such as watches with heart-rate monitors, oxygen sensors, position sensors, airflow sensors, electrocardiogram and also skin resistance sensors, which give you a very good idea of an individual's physiological state. They work equally well for people who are ill and for the worried well, also known as 'quantified selfers'.

Another iconic Internet of Things product is Nest, producing a machine-learning thermostat for the home that has just been bought by Google for three billion dollars.

I would be amiss in talking about machine learning and the revolution of big data without mentioning Solexa, the human-genome-sequencing company. It was founded by two Cambridge professors from the Chemistry lab, Shankar Balasubramanian and David Klenerman. Their technology brought down the cost of sequencing a human genome from 10 million dollars to 1,000 dollars (Figure 8.8). This cost reduction has continued, and the cost is now under 1,000 dollars with Illumina, the company that bought Solexa.

In comparison, look at the graph representing Moore's law. Genome sequencing tracked Moore's law for a long time, but the invention of sequencing by synthesis by Solexa produced the most dramatic decrease in cost of any important parameter I've ever come across in any sector: a factor of 10,000 in less than seven years.

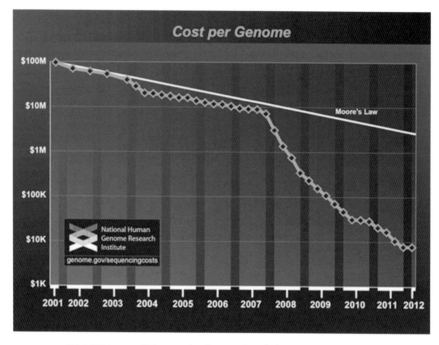

FIGURE 8.8 Solexa technology reduced the cost per genome from $10 million to less than $1,000. (Courtesy: National Human Genome Research Institute.)

What we have seen so far are all the different pieces of hardware that characterise a wave, but on top of that hardware we have seen a spectacular increase in the quality of software, both in terms of the size of the software programs and in terms of the ease of use and the way computer interfaces become more human-friendly, a good example being voice recognition.

The next revolution in software is machine learning. It is arguably the most exciting and powerful development in computer science, and it is quite a fundamental shift from the way we have been producing software in the past.

Everybody knows that computers work with zeros and ones, true or false, either something is that way or it is not. This has now been replaced by probabilities, which are much more appropriate to describe the real world.

In the past programmers have worked extremely hard to make computers error-free. Programs incorporate high levels of error correction so that you can have an incredibly high degree of confidence that the data that you get off your disk, for example, has no errors. There are even layers of error correction to make sure there are no errors.

We are now giving up this deterministic view of computers in favour of a statistical view of the world, which again is much more realistic, much more powerful.

In the future, we will not program computers anymore, instead we will teach them. The problem with teaching a computer is that it is a bit like teaching a child: he or she may do what you taught them or they may not. What is needed to make machine learning work well is big data; very big, high-quality training data sets.

But the biggest problem of machine learning is that just having the big training set is not enough. You also need to tell the machine what is a good thing and what is a bad thing, and that is a value judgement. It requires a human to provide ultimate goals of what to optimise, and this causes the genie problem. These programs will achieve exactly the goal set for them, just as the genie grants wishes literally, and humans are very bad at wishing for the right thing.

The user interface in machine learning is becoming much closer to a human user interface: so, for example, thermostats learn by themselves what temperature you prefer. There is a revolution about to break in healthcare too, with bio sensors that provide information to you 24/7 combined with the gene-association studies that we are now beginning to do with the 100,000 Genomes Project that David Cameron announced in 2014. Amadeus has just made an investment in a company called Congenica that can interpret the results of this genome project for rare diseases.

Let me touch upon a few very exciting developments in computing. The first is by Professor Steve Furber, one of the two people who designed the ARM processor at Acorn computers. He is using a million of these ARMs to simulate a billion neurones in real time as part of the Human Brain Project, which is a billion-euro project of the EU. Every single chip that he has developed actually contains 18 ARMs.

One cannot talk about technology development without mentioning quantum computing. Instead of bits of zeros and ones, qubits (quantum bits) take advantage of a peculiar phenomenon in quantum computing whereby they can be a combination of zero and one at the same time. Now this turns out to be very useful, because qubits can combine zero and one in many different ways, which means that they can carry much more information than a bit. If qubits are combined using another quantum phenomenon called entanglement, the processing power of these computers increases exponentially. There is now a record of 14 entangled qubits, and one only needs about 50 entangled qubits for quantum computers to be more powerful than traditional computers.

To return to the clash of waves: why does the new wave always win? The answer is higher volume, lower price, and the fact that the new wave soon overtakes the old wave in performance. New applications and the user interface are becoming more humanlike. A typical example is the PC being replaced by mobile phones, which are a lot easier to use because you can talk to them.

ARM versus Intel is another case in point, RISC vs. CISC, Reduced Instruction Set Computer architecture vs. Complex Instruction Set Computer architecture. Here is the story.

When we needed a new processor at Acorn in the early 1980s we went to Intel because we wanted to use their 80286 processor. We told them that we thought their processor was quite interesting, but they had a bad pinout because both the data bus and the address bus were on the same pins, and nobody could make a sensible computer out of that. We asked them to sell us the chip itself and we would do our own pinout and maybe make something of the chip. Acorn at that time was a very successful Cambridge start-up, and we were quite cocky, but they said 'Get lost.'

With the confidence of youth we said 'You get lost, we'll do our own chip.' This is the only reason why ARM exists; if they had given us that processor, they would not have the problem with ARM that they do now.

It is also an interesting example of a reverse set of roles between the USA and the UK. Normally the UK invents products and the USA exploits them, but RISC was invented by John Hennessy and David Patterson in Stanford and Berkeley, respectively. The first commercial implementation of these ideas, however, was done in Cambridge at Acorn. Originally ARM stood for Acorn RISC Machine. We only changed it to Advanced RISC Machine when we spun it out with Apple.

The ARM had 20 times the performance of the Z80, which was the most popular microprocessor at the time, but the Z80 and the ARM used exactly the same number of transistors. This showed just how potent that new idea of RISC was and how smart Steve Furber and Sophie Wilson were in terms of creating the new architecture.

I gave two advantages to the ARM design team that neither Intel nor Motorola, nor any of the other processor firms, ever managed to give to their design teams. They were as follows:

1. I gave them no people: it is the only processor that was ever designed by just two people.
2. I gave them no money, because we did not have any. So there was no way they could produce a complex chip.

This became a very, very simple chip, but not one of low performance. It turned out to have very high performance, but a side effect of the simplicity was the low power consumption, which was not an original design goal. We became world record holders of MIPS per watt. Acorn did not need low power, as the BBC micro was mains powered, we just wanted to make sure it fits into a plastic package without a heat sink. But it turned out to be very important for Nokia because mobile phones are battery operated. ARM now has more than 95% market share in mobile phones, and we have now cumulatively sold over 70 billion ARMs; that is 10 ARMs per person on Earth.

You might ask 'Where are my 10 ARMs?' They are in your pocket: an iPhone has more than 10 ARMs. ARM has become our most successful Cambridge company, with a market capital of 25 billion dollars. But what is not very widely known is that since 2010 the value of the ARM chips that are produced by our licensees has overtaken Intel sales. So even in dollar terms the ARM architecture is now more important than Intel, and the fight between Intel and ARM is no longer just between Intel and ARM. It is the fight between Intel and the rest of the world, represented by the 400 ARM licensees, companies like Qualcomm, Toshiba, Sony, Lenovo, Apple. In fact, every single sizable semiconductor company in the world, including Intel, has an ARM licence.

I would like to conclude with some remarks about Alan Turing, the present Artificial Intelligence (AI) hype, the risks associated with AI, the

159

state of the art, and the dangers of superintelligences that we might not be able to handle properly.

Alan Turing, in his paper on machine intelligence in 1950, asked the question 'Can computers think?' Rather than answer the question directly, he answered it with a test, known as the Turing test: if you talk to a computer and you cannot tell whether it is a computer or a human answering, the computer is intelligent.

Artificial Intelligence is commonly perceived through movies, with probably the most famous being the Schwarzenegger movie *Terminator*. But there are subtler movies now, like *Her*, *Ex Machina*, and *Transcendence*, that deal with the subject of AI in a more thoughtful way. There are also a number of books like Nick Bostrom's book on superintelligence and the new book by Murray Shanahan on *The Technological Singularity*, both worth reading.

Lord Rees in 2003 published *Our Final Century*, sold in the USA as *Our Final Hour*, because Americans want instant gratification. He noticed very early on that these unlikely but extreme events including AI disasters can and will happen unless we are very careful. More recently, Stephen Hawking has said that 'success in creating AI would be the biggest event in human history, unfortunately it might also be the last unless we learn how to avoid the risks', and Elon Musk, the creator of the Tesla car and SpaceX, said 'I hope we are not just the biological boot loader for digital superintelligence. Unfortunately that is increasingly probable.'

There are a number of organisations looking at these problems. The Centre for the Study of Existential Risk (CSER) in Cambridge, founded by Martin Rees and Huw Price, holds workshops and meetings that I can highly recommend. Also the Future of Humanity Institute in Oxford with Professor Nick Bostrom and the Future of Life Institute in Boston, the other Cambridge across the pond, which Elon Musk is supporting, focus on these issues.

So where are we? How intelligent are these computers? Well, we have had a computer chess champion with Deep Blue since 1997. We have IBM Watson, which beat the two human champions in 2011 to become *Jeopardy* champion (that is an American quiz show), and with excellent speech recognition, we've got good natural understanding, not just of

English but also of German, Spanish, Portuguese, Chinese, and Japanese. Computers now recognise them all. This is one of the great advantages of a scalable architecture, that all the major languages are now recognised by machines.

We now have encyclopaedic knowledge through Wikipedia. Geoffrey Hinton produced a revolutionary paper on object recognition in 2014. Computer face recognition is now better than human. How did that happen? Because the computer was trained on 15 million faces. Now we are pretty good at remembering faces, but at 15 million we have to give up while a computer does not.

There are self-driving cars with impressive performance in normal traffic. Computer diagnosis of tumours is now better than by the best pathologists. These are examples of where AI is really becoming very useful. So, what is missing? The answer is common sense.

Computers are not so good at this yet. Neither are computers very good at creativity, although in my opinion creativity in the final analysis is really just a very fast random ideas generator with a very efficient filter. AI does not have human values, and it cannot make breakfast. But there is agreement in the AI community that Artificial General Intelligence (AGI) is possible, and that we will have such superintelligences by about 2050.

So the key question becomes 'How can we ensure that AGI is not just good at chess-playing or self-driving cars, but at general intelligence, like a human intelligence embodying human values?' The answer is that nobody knows. I think it is a key question that we should all take very seriously.

More than 50% of jobs are under threat from these superintelligences. It may happen very fast. The associated productivity increase will probably mean that the increased unemployment will not lead to poverty, as there will be enough money in the new economy to support everyone. The question is whether people will be able to adjust to not working or changing to a different job fast enough.

In conclusion, I have discussed general-purpose technologies of which there have been 24, although I would suggest artificial intelligence is the 25th for the twenty-first century. I discussed the six waves of computing and how one supersedes the other, the clash between the waves, the nice story of a local company ARM giving Intel a hard time and Artificial General Intelligence, but I will end with a plea: it took us 70 years after the car was invented to provide safety belts. We will not have 70 years between now and superintelligences to prevent bad things from happening. I would urge you to engage in the debate about artificial intelligence. Read books on the subject, come to CSER meetings, inform yourselves, because if we get this wrong the consequences could be much worse than a few car crashes.

Index

Index